Lecture Notes in Computer Science   15189

Founding Editors

Gerhard Goos
Juris Hartmanis

W0037493

The series Lecture Notes in Computer Science (LNCS), including its subseries Lecture Notes in Artificial Intelligence (LNAI) and Lecture Notes in Bioinformatics (LNBI), has established itself as a medium for the publication of new developments in computer science and information technology research, teaching, and education.

LNCS enjoys close cooperation with the computer science R & D community, the series counts many renowned academics among its volume editors and paper authors, and collaborates with prestigious societies. Its mission is to serve this international community by providing an invaluable service, mainly focused on the publication of conference and workshop proceedings and postproceedings. LNCS commenced publication in 1973.

Diego Marmsoler · Meng Sun
Editors

Formal Aspects of Component Software

20th International Conference, FACS 2024
Milan, Italy, September 9–10, 2024
Proceedings

 Springer

Editors
Diego Marmsoler 🆔
University of Exeter
Exeter, UK

Meng Sun 🆔
Peking University
Beijing, China

ISSN 0302-9743   ISSN 1611-3349 (electronic)
Lecture Notes in Computer Science
ISBN 978-3-031-71260-9   ISBN 978-3-031-71261-6 (eBook)
https://doi.org/10.1007/978-3-031-71261-6

Preface

This volume contains the proceedings of the 20th International Conference on Formal Aspects of Component Software (FACS 2024), held in Milan, Italy during September 9–10, 2024.

FACS aims to bring together practitioners and researchers in the areas of component software and formal methods to promote a deeper understanding of how formal methods can or should be used to make component-based software development succeed. The component-based software development approach has emerged as a promising paradigm to transport sound production and engineering principles into software engineering and to cope with the ever-increasing complexity of present-day software solutions. However, the advent of emerging computational paradigms, which requires better support from component-based solutions, e.g., cyber-physical human systems, quantum computations, AI systems and blockchain, has posed many new challenging research questions, which require established concepts and techniques to be revisited and new ones to be developed in order to meet the opportunities offered by these challenges.

FACS 2024 is the 20th International Conference on Formal Aspects of Component Software, a series of symposia started in 2003. In past years, FACS took place in Pisa (2003), Macau (2005), Prague (2006), Sophia Antipolis (2007), Málaga (2008), Eindhoven (2009), Guimarães (2010), Oslo (2011), Mountain View (2012), Nanchang (2013), Bertinoro (2014), Rio de Janeiro (2015), Besançon (2016), Braga (2017), Pohang (2018), Amsterdam (2019), and online (2021–2023). This year we received 16 submissions covering different areas of formal aspects of component software. Each paper was reviewed by at least three reviewers and the Program Committee accepted 7 regular long papers and 1 short paper, leading to an attractive scientific program. Authors of selected accepted papers will be invited to submit extended versions of their contributions to appear in a special issue of Elsevier's Science of Computer Programming journal.

FACS 2024 would not have succeeded without the deep investment and involvement of the Program Committee members and the external reviewers who evaluated and selected the best contributions. We would like to express our gratitude to all the authors who submitted their work to the conference, the Steering Committee members who provided precious guidance and support, all the colleagues who served on the Program Committee, as well as the external reviewers, whose professional and efficient work during the review process helped us to produce a high-quality conference program. Particular thanks are given to the invited speakers, Ana Cavalcanti from the University of York, David Parker from the University of Oxford, and Geguang Pu from East China Normal University, for their willingness to talk about their research and share their perspective about formal aspects of component software. The abstracts of the invited talks are included in this volume as well.

The EasyChair system was set up for the management of FACS 2024, supporting submission, review, and volume preparation processes. It proved to be a powerful framework. We thank Springer for the smooth cooperation in the production of this proceedings volume.

FACS 2024 was co-located with Formal Methods 2024, hosted and sponsored by Politecnico di Milano, Italy. The local organization committee offered all the facilities to run the event in a lovely and friendly atmosphere. Many thanks to all the local organizers.

July 2024   Diego Marmsoler
                                                                            Meng Sun

Organization

Program Committee Chairs

Diego Marmsoler	University of Exeter, UK
Meng Sun	Peking University, China

Steering Committee

Farhad Arbab	CWI and Leiden University, Netherlands
Kyungmin Bae	Pohang University of Science and Technology, South Korea
Peter Csaba Ölveczky	University of Oslo, Norway
Javier Cámara	University of Málaga, Spain/University of York, UK
Sung-Shik Jongmans	Open University and CWI, Netherlands
Zhiming Liu	Southwest University, China
Markus Lumpe	Swinburne University of Technology, Australia
Eric Madelaine	Inria Sophia Antipolis, France
Corina Pasareanu	CMU, USA
José Proença	Polytechnic Institute of Porto, Portugal
Gwen Salaün	Université Grenoble Alpes, France
Luís Soares Barbosa	University of Minho, Portugal
Anton Wijs	Eindhoven University of Technology, Netherlands

Program Committee

Giorgio Audrito	University of Turin, Italy
Kyungmin Bae	POSTECH, South Korea
Luís Soares Barbosa	University of Minho, Portugal
Simon Bliudze	Inria Lille, France
Achim Brucker	University of Exeter, UK
Zhenbang Chen	NUDT, China
Brijesh Dongol	University of Surrey, UK
Clemens Dubslaff	Eindhoven University of Technology, Netherlands
Marie Farrell	University of Manchester, UK
Simon Foster	University of York, UK

Samir Genaim	Universidad Complutense de Madrid, Spain
Fatemeh Ghassemi	University of Tehran, Iran
Mario Gleirscher	Universität Bremen, Germany
Keigo Imai	DeNA Co., Japan
Kenneth Johnson	Auckland University of Technology, New Zealand
Violet Ka I Pun	Western Norway University of Applied Sciences, Norway
Olga Kouchnarenko	University of Franche-Comté, France
Ivan Lanese	University of Bologna/Inria, Italy
Antónia Lopes	Universidade de Lisboa, Portugal
Diego Marmsoler	University of Exeter, UK
Mieke Massink	CNR-ISTI, Italy
Jacopo Mauro	University of Southern Denmark, Denmark
Peter Ölveczky	University of Oslo, Norway
José Proença	University of Porto, Portugal
Camilo Rocha	Pontificia Universidad Javeriana Cali, Colombia
Gwen Salaün	University of Grenoble Alpes, France
Arpit Sharma	IISERB, India
Meng Sun	Peking University, China
Anton Wijs	Eindhoven University of Technology, Netherlands
Shoji Yuen	Nagoya University, Japan
Min Zhang	ECNU, China
Xiyue Zhang	University of Oxford, UK

Additional Reviewers

Frederic Dadeau
Weijiang Hong

Keynotes

Comparing Reactive Models and Cyclic Components of Robotic Systems: The RoboStar Approach to Model-Based Testing

Ana Cavalcanti

University of York, UK

We present support for automatic generation of simulation tests based on reactive models described in RoboChart, a domain-specific notation for verification of reactive designs of control software for robotic systems. The soundness and completeness of our approach are established using RoboChart's process algebraic semantics, formalised in tock-CSP, a discrete-time variant of CSP.

RoboChart employs state machines enriched with time constructs and a component model to specify timed, platform-independent behavioural models. The RoboChart semantics defines processes that characterise the reactive behaviour of robotic control software through interactions representing sensor outputs and actuator inputs. Support for modelling, model checking, and test generation based on RoboChart is available via RoboTool[1]. RoboChart is part of the RoboStar framework for design and verification of mobile and autonomous robots.

The testing theory for tock-CSP involves constructing tests from forbidden traces of a process, including input and output events and a special event, namely, *tock*, representing the passage of time. RoboTool automates the process of generating forbidden traces and constructing tests from them.

Simulations are widely used in robotics for early testing. While RoboChart models are reactive, however, simulations often follow an idealised cyclic paradigm. In each cycle, the simulation under test (SUT) evolves by accepting inputs, processing them, and providing outputs, infinitely fast, before advancing time to the next cycle. Our work with RoboChart involves comparing reactive and cyclic models using a conformance relation defined and formalised in tock-CSP. Test generation and execution must adapt to this notion of conformance.

In our work, we identify the reactive tests based on forbidden traces that are meaningful, define a process to convert those tests, provide an algorithm to execute the tests, and prove the soundness and completeness of our approach. The testing approach we present is a significant advancement in current testing practices within the field of robotics, where simulations are widely used.

---

[1] robostar.cs.york.ac.uk/robotool/.

Acknowledgement. This work is joint with Rob H. Hierons. The author would like to thank the RoboStar team for useful discussions. This work is funded by the Royal Academy of Engineering under Grant No CiET1718/45, the UKRI (UK Research and Innovation Council) under Grants No EP/R025479/1 and EP/V026801/1, and by EU Horizon project RoboSapiens under agreement number 101133807.

Verification and Control of Stochastic Multi-agent Systems

David Parker ⓘ

Department of Computer Science, University of Oxford, Oxford, UK
`david.parker@cs.ox.ac.uk`

Probabilistic model checking provides a powerful set of techniques for automated formal verification of computerised systems operating in uncertain environments. Extensions of these methods to multi-agent systems open up the possibility to formally model and analyse systems with multiple components or actors, potentially acting in either an adversarial or collaborative manner. This talk surveys some recent advances in this area, bringing together techniques and tools from formal verification with game-theoretic concepts and methods, notably stochastic games and equilibria. We illustrate applications of this approach to component-based design and analysis in a variety of application domains, from multi-robot control to distributed protocols for network communication or energy management. We also discuss some of the key challenges and research directions to make further progress in the area.

The Experiences of Developing the Industrial-Strength Tools for Modeling, Testing and Verification: A Formal Methods Perspective

Geguang Pu

East China Normal University, China

FM-related techniques are very helpful to ensure the quality of software. For instance, the model checking technique has been successfully applied in hardware/software verification and it becomes the key element for EDA tool chains. In this talk, I will share the experiences of developing the industrial-strength tools for modeling, testing and verification from the formal methods perspective. We will show how to find the real problems about testing and verification from the industry and also show how the formal methods can guide to solve these problems by developing tools. We will illustrate our experiences and insights by three interesting tools under development. The first one is a formal modeling and verification tool for the formal verification of Interlock system of the train, that is key part of signal systems in railway. The second one is a testing tool for embedded systems, where the symbolic execution technique plays an important role. The last one is a new model checking solver for hardware verification, and its performance is tuned effectively by the new observations on the search process in the state space. These tools are successfully applied in our industry partners and their effectiveness is also proved by large scale examples. Last but not least, we will also share the lessons we have learned during the tool development.

Contents

Verification and Testing

Enabling Behaviour Tree Verification
via a Translation to BIP

Qiang Wang[1(✉)], Huadong Dai[1], Yongxin Zhao[2], Min Zhang[2], and Simon Bliudze[3]

[1] Academy of Military Sciences, Beijing, China
18513688908@163.com
[2] East China Normal University, Shanghai, China
[3] Univ. Lille, Inria, CNRS, Centrale Lille, CRIStal, UMR 9189, 59000 Lille, France
simon.bliudze@inria.fr

Abstract. A formal verification method for behavior tree (BT) is proposed. The method is based on a compositional model transformation of BT into the formal component-based system design framework BIP (Behavior-Interaction-Priority). The transformation encodes each BT node as an individual BIP component that is formally defined by an extended finite state automaton (FSA), and each BT edge as a set of interactions that describes the allowed coordination between components. The correctness proof of the model transformation is presented, and a prototype tool-chain has been implemented that enables the automated verification of BT. Two practical case studies show that the tool-chain can not only verify the correctness of BT, but also detect the potential design flaws automatically.

Keywords: Behavior tree · Formal verification · BIP framework · Robotics

1 Introduction

Behavior tree (BT) is becoming increasingly popular in robotics and autonomous systems as a control architecture for task representation and execution [12, 16]. Due to the fact that BT is highly flexible, modular and well suited for defining deliberative elements in model-based design of robotics software, they have been adopted in both academic and industrial settings. For instance, BT has been used in the Boston Dynamics Spot SDK for modeling the robot's mission, and in the Navigation and Task Planning System of ROS2, the *de facto* standard in robotics software development.

Given the conceptual simplicity, BT has been mostly presented in an informal way (e.g., pseudocode [11]), and their implementations in different libraries are usually subject to different interpretations, whose correctness with respect to essential properties (e.g., safety properties) has not been formally verified. Nevertheless, as noted in several related works [3,9,17], when applying BT in safety-critical systems, it is indispensably important to rigorously verify and guarantee their safety. To this end, this work is concerned with the following problem: given a BT and a property specification, how can we check whether the BT satisfies the given property in an automated manner?

This work presents a new method for the formal verification of BT. Specifically, for the formal modeling of BT, we leverage on the BIP framework, which is a component-based system design framework for safety-critical systems, e.g., autonomous robots

D. Marmsoler and M. Sun (Eds.): FACS 2024, LNCS 15189, pp. 3–20, 2024.
https://doi.org/10.1007/978-3-031-71261-6_1

[1,2,5,18]. A compositional model transformation from BT to BIP has been presented, where each BT node is transformed into an individual BIP component as a FSA extended with real-valued data and communication ports, and each BT edge is transformed into a set of interactions, describing the allowed coordination between components, e.g., passing a tick to a child, or returning a status to the parent. The formal BIP model of the entire BT is then composed by assembling all the components using interactions. A key advantage of this transformation is that all BT nodes are handled individually, and all nodes with the same type share a common component interface, which maintains the modularity and flexibility of BT. Last but not least, the correctness of the model transformation has been proven based on the formal semantics of BIP, and a tool-chain has been implemented that integrates both the model transformation and the BIP model checking tools. This tool-chain enables us to verify the correctness of BT with respect to essential safety property specifications in an automated manner. Two practical case studies demonstrate the usefulness and effectiveness of the proposed method and tool-chain.

We remark that the proposed method overcomes several deficiencies of the other related work, e.g., [3]. On the one hand, the proposed method can handle computations in the BT nodes, e.g., arithmetic over integer variables, and thus it is able to model and analyze infinite state models. While the work in [3] represents all BT nodes symbolically without taking into account their computations, and thus can only handle finite state models. On the other hand, the proposed method enables automated verification of BT. While the work in [3] still relies on a manual encoding of BT into linear temporal logic (LTL) [8] for formal verification. Finally, although FSA-based model checking has been successfully used for the automated formal verification of safety-critical systems [8], it has not been widely used for BT. A plausible reason as reported in [3] may be that classical model checking methods based on FSA usually constructs a transition system that requires low-level knowledge of behavior in every system state, violating modularity and deconstructing the tree structure that gives BT the flexibility and reusability. However, in this work we show that with a proper compositional transformation of BT into the FSA-like representation, model checking techniques can still be used without compromising the modularity and flexibility of BT.

The rest of the paper is organized as follows. Section 2 reviews the most related work on formal analysis of BT. Section 3 presents a brief introduction to BT and the BIP framework. Section 4 and Sect. 5 present a formal model for BT and the transformation into the BIP framework, respectively. Section 6 presents the proposed formal verification technique of BT. Section 7 concludes the paper with some future work.

2   Related Work

We review the most related work on the formal analysis of BT, and refer to the survey [16] for a detailed introduction of their applications into various domains. A formal verification framework based on LTL has been proposed in [3]. The framework encodes both BT and the property as logical formulae in LTL and reduces the verification problem to the LTL satisfiability problem. The work in [9] formalizes BT and its execution context using program graphs and applies runtime monitoring to check whether the system behaves correctly under the control of the given BT. More recently, the work in [15] presents a formal verification method based on an encoding of BT into Horn clauses. The work in [21] presents a formal verification method by transforming BT to nuxmv models [6]. Instead of verifying the correctness of BT, the work in [10] considers the synthesis of BT in a correct-by-construction manner, and proposes an approach to generate BT that are guaranteed to satisfy the desired properties specified in a fragment of LTL. A formal semantics for BT has been presented in [14] by encoding BT into the communicating sequential processes (CSP), that is a formal language for describing interaction patterns in concurrent systems. However, the considered BT mainly represents a requirement modelling language, which is different from the control architecture for robotic systems studied in this work.

Finally, we remark that for the design of robotic systems, some related work proposes to combine UML/SysML with BIP [7,20], where the system behavior is modeled using UML/SysML and then translated into a formal specification in BIP for correctness verification. In this work we focus on the transformation from BT to BIP for formal verification. This would also be beneficial to the robotics design in our opinion, since BT is expressive and convenient enough to represent task and control structures to be coded in robotic systems. Notably, the work in [13] shows that BT generalizes several related control structures based on a functional representation of BT. The work in [19] investigates the equivalence between BT and the controlled hybrid dynamical systems. Furthermore, the work in [4] shows the expressiveness of BT compared to other control architectures using a common abstract framework, called action selection mechanism.

3   Preliminaries

3.1   A Brief Introduction to BT

A BT is a tree-structured control architecture, where the leaf nodes represent the computation tasks to be executed or the conditions to be checked, while the control nodes orchestrate the flow of task executions. The classical formulation of BT in [11,13] introduces two types of leaf nodes, i.e., action and condition nodes, and three types of internal control nodes, i.e., sequence, fallback (a.k.a. selector) and parallel nodes. The execution of a BT starts from the unique root node, which continuously sends activation signals (a.k.a. ticks) to its children in the DFS manner with a given frequency. When a control node is ticked, it passes the tick to the leftmost child first and decides whether the passing continues to the right children or return to its parent according to the returned status of the ticked child. When a leaf node is ticked, it starts the execution and returns a status to its parent upon completion, which can be either *Success*, *Failure* or *Running*. Table 1 summarizes the conditions under which each status is returned for the five node types.

The action node returns *Success* if the task is completed successfully, *Running* if the task is still ongoing, and *Failure* otherwise. The condition node returns *Success* if the condition being checked is satisfied, and *Failure* otherwise. It never returns a *Running* status. When the sequence node receives a tick signal, it starts ticking its children from left to right sequentially and waits for the returned status from each child. If one child returns a *Failure* or *Running* status, the sequence node stops ticking the next child (if any) and returns the *Failure* or *Running* status, respectively. It returns *Success* only if all the children return *Success*. The fallback node reacts similarly on receiving a tick. It returns a *Success* or *Running* status and stops ticking the subsequent children if a child returns *Success* or *Running*, respectively. And it returns *Failure* if all the children return *Failure*. The parallel node is parameterised with a success threshold M. When receiving a tick, the parallel node passes the tick to all the children, and returns *Success* if M out of n children return *Success*, *Failure* if more than $n - M$ children return *Failure*, and *Running* otherwise.

Table 1. The condition of each status for the five BT node types

Node type	Symbol	Success	Failure	Running
Action	box	Upon completion	If impossible to complete	During execution
Condition	oval	If satisfied	If unsatisfied	Never
Sequence	→	If all children succeed	If one child fails	If one child returns running
Fallback	?	If one child succeeds	If all children fail	If one child returns running
Parallel (with a parameter M)	⇒	If at least M children succeed	If more than $N - M$ children fail	else

Example 1. Figure 1 shows an example BT that controls the movements of a robot [13]. It contains three control nodes and four leaf nodes, and implements the following control policy. When executing the BT, the battery level condition is checked, and if the condition holds (i.e., the battery level is above 20%), the *Do other task* action node is activated, which might issue some commands to the robot actuators. Otherwise, the action nodes *Move to station* and *Recharge battery* are executed sequentially to drive the robot to the station for recharging. Overall, the execution repeats from the top-level sequence node by ticking the children nodes periodically. The desired safety property is that the controlled robot will never run out of battery.

3.2   The BIP Framework

The BIP framework provides a language with well-defined semantics for component-based system modeling and an associated toolset to support the correct-by-construction design. The BIP toolset includes a model checker for the verification of essential safety properties and a code generator that produces C/C++ codes from BIP models. Following our previous work on formal verification of BIP models [5, 18], we take the fragment of BIP with multiparty synchronization and data transfer for the encoding of BT. For the sake of space, we briefly introduce the key concepts of the BIP modeling language and refer to [5] for more details.

Given a set of real-valued variables $\mathcal{V}$, let $\mathcal{E}_{\mathcal{V}}$ and $\mathcal{F}_{\mathcal{V}}$ represent the set of boolean expressions and the set of functions over $\mathcal{V}$, respectively. A BIP model consists of a finite set of components, each of which is described by a finite state automaton extended with data and ports. Formally, a BIP component is defined as follows.

Definition 1 (Atomic component]). *An atomic BIP component is defined as a tuple* $\mathcal{B} = \langle \mathcal{V}, \mathcal{L}, \mathcal{P}, \mathcal{T}, \ell \rangle$, *where*

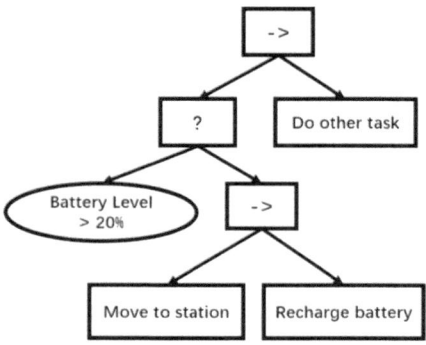

Fig. 1. An example BT that controls the movements of a robot.

1. $\mathcal{V}$ *is a finite set of state variables;*
2. $\mathcal{L}$ *is a finite set of control locations;*
3. $\mathcal{P}$ *is a finite set of communication ports;*
4. $\mathcal{T} \subseteq \mathcal{L} \times \mathcal{P} \times \mathcal{E}_\mathcal{V} \times \mathcal{F}_\mathcal{V} \times \mathcal{L}$ *is a finite set of transitions;*
5. $\ell \in \mathcal{L}$ *is the initial control location.*

Each transition of the component can be labelled by a port in $\mathcal{P}$, a boolean expression in $\mathcal{E}_\mathcal{V}$ representing the enabling guard and a function in $\mathcal{F}_\mathcal{V}$ representing the computation task. When no computation task is specified in the transition, we denote it by *null*. A transition can be taken if the labelling guard is enabled by the current state, and the associated function is executed when the transition is taken. If multiple transitions are enabled simultaneously, one of them is chosen non-deterministically. Moreover, every pair of outgoing transitions from the same control location have different ports, i.e., for each pair of transitions $(l_1, p_1, e_1, f_1, l'_1) \in \mathcal{T}$ and $(l_2, p_2, e_2, f_2, l'_2) \in \mathcal{T}$, if $l_1 = l_2$, then $p_1 \neq p_2$.

The set of ports $\mathcal{P}$ forms the interface of a component and they are used to define the interaction model between different components. In BIP, the interaction model provide the means to compose a finite set of components. Formally, an interaction model is defined as follows.

Definition 2 (Interaction model). *An interaction model Γ is a finite set of interactions, each of which is formally defined by a tuple $\gamma = \langle g, \mathcal{P}, f \rangle$, where*

1. $g \in \mathcal{E}_\mathcal{V}$ *is an enabling guard;*
2. $\mathcal{P} \subseteq \bigcup_{i=1}^{n} \mathcal{P}_i$ *is a non-empty set of ports from n components and $\forall i \in [1, n]$, $|\mathcal{P} \cap \mathcal{P}_i| \leq 1;$*
3. $f \in \mathcal{F}_\mathcal{V}$ *is a function representing message passing.*

Intuitively, an interaction model defines a guarded multi-party synchronization with data transfer: when the guard g of an interaction γ is enabled, then function f will be executed, and all the transitions labelled by ports in $\mathcal{P}$ are taken synchronously. The constraint $\forall i \in [1, n], |\mathcal{P} \cap \mathcal{P}_i| \leq 1$ ensures that for each component, there can be at most one port per interaction. In other words, it is not allowed to involve two ports from the same automaton in a single interaction, since it makes no sense to synchronize an automaton with itself. For simplicity, we use notation $\mathcal{B}.p$ to denote the port p of component $\mathcal{B}$.

Given a set of interactions Γ and a set of components $\{\mathcal{B}_i \mid i \in [1, n]\}$, a BIP model is formally defined by a tuple $\mathcal{M} = \langle \{\mathcal{B}_i \mid i \in [1, n]\}, \Gamma \rangle$.

4   A Formal Model of BT

Let $NodeType = \{C, A, S, F, P\}$ denote the BT node types as shown in Table 1. We introduce a formal definition of the BT node as follows.

Definition 3 (BT node). *A BT node is defined by a tuple* $\mathcal{N} = \langle \mathcal{V}_d, \mathcal{V}_n, t, g_s, g_f, g_r, f \rangle$, *where*

1. $\mathcal{V}_d$ *is a set of data variables;*
2. $\mathcal{V}_n$ *is a set of node variables;*
3. $t \in NodeType$ *is the type of node;*
4. $g_s, g_f, g_r \in \mathcal{E}_{\mathcal{V}_d \cup \mathcal{C}_n}$ *are boolean expressions representing the Success, Failure and Running conditions, respectively;*
5. $f \in \mathcal{F}_{\mathcal{V}_d}$ *is a function representing the computation task.*

Given a node $\mathcal{N}$, we use notation $\mathcal{N}.y$ to denote the element of $\mathcal{N}$, where $y \in \{\mathcal{V}_d, \mathcal{V}_n, t, g_s, g_f, g_r, f\}$ for simplicity. We denote by $dom(x)$ the domain of a variable x. For a node variable $x \in \mathcal{V}_n$, its domain $dom(x)$ is the set of children nodes of $\mathcal{N}$. Moreover, the function f of a condition node is always null and the running condition g_r is always false, since it never performs a computation task. We assume that all action nodes are finite time successful [12].

Intuitively, a BT is a tree-like structure, where each vertex is associated with a BT node. Its formal definition is given as follows.

Definition 4 (BT). *A BT is formally defined by a tuple* $\mathcal{BT} = \langle \mathcal{NS}, V, E, L, v_0 \rangle$, *where*

1. $\mathcal{NS}$ *is a set of BT nodes;*
2. V *is a set of vertices;*
3. $E \subseteq V \times V$ *is a set of edges;*
4. $L : V \rightarrow \mathcal{NS}$ *is a labeling function that assigns a BT node* $\mathcal{N} \in \mathcal{NS}$ *to a vertex* $v \in V$;
5. $v_0 \in V$ *is a unique root vertex.*

We say that vertex v' is a child of vertex v if there is an edge $e \in E$ such that $e = (v, v')$. For simplicity, for edge $e = (v, v')$, we denote by $source(e) = v$ and $target(e) = v'$. The set of children of vertex v are then denoted by $children(v) = \{target(e) \mid e \in E \wedge source(e) = v\}$. Similarly, the parent node of v is denoted by $parent(v) = \{source(e) \mid e \in E \wedge target(e) = v\}$.

A BT $\mathcal{BT} = \langle \mathcal{NS}, V, E, L, v_0 \rangle$ is well-defined if the following conditions hold:

1. The set of vertices and edges $\langle V, E \rangle$ forms a tree, and $\forall v, v' \in V.\ v \neq v'$, then $L(v) \neq L(v')$;
2. There is an edge $(v, v') \in E$ if and only if the output of BT node $L(v)$ is an input to BT node $L(v')$;
3. For each vertex $v \in V$, we have $|children(v)| \geq 0$. Moreover, when $|children(v)| > 0$, $children(v)$ is an ordered set;
4. For each vertex $v \in V$ such that $|children(v)| = 0$, we have $L(v) = \mathcal{N}$ where $\mathcal{N}.t \in \{C, A\}$ and $|\mathcal{N}.\mathcal{V}_n| = 0$;
5. For each vertex $v \in V$ such that $|children(v)| \neq 0$, we have $L(v) = \mathcal{N}$ where $\mathcal{N}.t \in \{S, F, P\}$ and $|\mathcal{N}.\mathcal{V}_n| = |children(v)|$;
6. For each vertex $v \in V$, we have $|parent(v)| \leq 1$. Moreover, if $|parent(v)| = 0$, then $v = v_0$ and $L(v)$ is the root BT node.

Example 2 Considering the BT in Fig. 1, it is defined by a tree with seven vertices and six edges, where each vertex is labelled by a BT node. For the sake of space, we illustrate the formal definition using an action node and a sequence node. Let $x_1 \in [0, 100]$ and $x_2 \in [0, 1]$ be the state variables denoting the distance from the current position to the recharging station and the battery level, respectively. The action node *Do other task* is formally defined by a tuple $\mathcal{N}_1 = \langle \{x_1, x_2\}, \emptyset, A, false, false, true, f_1 \rangle$, where $f_1(x_1, x_2) = (x_1 + (50 - x_1)/50, x_2 - 0.1)$. While this action node is running, it issues commands to actuators to move the robot towards the position $x_1 = 50$, and makes the battery level decrease. Both success and failure conditions are false, i.e., it never returns a success or failure status, but always returns a running status. The sequence node is formally defined by a tuple $\mathcal{N}_2 = \langle \emptyset, \{y_1, y_2\}, S, g_s, g_f, g_r, null \rangle$, where y_1 and y_2 are node variables referring to the children nodes, and $g_s = y_1.g_s \wedge y_2.g_s$, $g_f = y_1.g_f \vee (y_1.g_s \wedge y_2.g_f)$, $g_r = y_1.g_r \vee (y_1.g_s \wedge y_2.g_r)$. In other words, a sequence node will return immediately with a status code failure or running when one of its children returns failure or running. It is often used when some actions are meant to be carried out in sequence, and when the success of one action is needed for the execution of the next.

We remark that the condition under which a status is returned by an action node is user-defined. Although the *Do other task* action node always returns a running status, it does not mean that the execution of the BT is blocked and waiting for the completion of this action node. Instead, the root node will restart ticking its children nodes (from the leftmost) once receiving the running status. Since the sequence node will find and execute the first child that does not return success, the two other actions nodes (i.e., *Move to station* and *Recharge battery*) start executing as soon as the *Do other task* action node brings the battery level below 20%.

5 Transforming BT to BIP

Given a $\mathcal{BT} = \langle \mathcal{NS}, V, E, L, v_0 \rangle$, the transformation of $\mathcal{BT}$ into a BIP model consists of two steps. The first step is to transform each BT node $\mathcal{N} \in \mathcal{NS}$ into a BIP component $\mathcal{B_N}$, and the second step is to transform the edges between BT nodes into a BIP interaction model $\Gamma_{\mathcal{BT}}$. Then by combing these two parts, we obtain the complete BIP model for $\mathcal{BT}$ $\mathcal{M}_{\mathcal{BT}} = \langle \{\mathcal{B_N} \mid \mathcal{N} \in \mathcal{NS}\}, \Gamma_{\mathcal{BT}} \rangle$.

5.1   Transformation of BT Nodes

Condition Node. The transformation for this type of node is rather straightforward. Let $\mathcal{N} = \langle \mathcal{V}_d, \mathcal{V}_n, C, g_s, g_f, g_r, f \rangle$ be a condition node, it is transformed into a BIP component $\mathcal{B}_{\mathcal{N}} = \langle \mathcal{V}', \mathcal{L}, \mathcal{P}, \mathcal{T}, \ell \rangle$, where

- $\mathcal{V}' = \mathcal{V}_d$;
- $\mathcal{L} = \{l_0, l_1\}$;
- $\mathcal{P} = \{tick, success, failure, running\}$;
- $\mathcal{T} = \{(l_0, tick, true, null, l_1), (l_1, success, g_s, null, l_0),$
  $(l_1, failure, g_f, null, l_0)\}$;
- $\ell = l_0$.

A graphical representation of this component is shown in the left part of Fig. 2. Recall that the set of ports $\mathcal{P}$ forms the interface of a component and they are used to define the coordination between different nodes. For each condition node, the corresponding component is equipped with four ports for communicating with its parent, where ports *success* and *failure* are used to return the corresponding status, and port *tick* is used to receive the ticking signal. Note that the port *running* is never used in the automaton, since a condition node never returns a running status. However, it is introduced in the formulation in order to have a uniform interface for all types of leaf nodes.

The automaton of this component consists of two control locations l_0 and l_1 with l_0 being the initial location. From the initial location, the transition labelled by port *tick* is taken on receiving a ticking signal, leading the component to location l_1. From location l_1, if the condition $[g_s]$ (or $[g_f]$) holds, the transition labelled by *success* (and *failure*, respectively) is taken and the corresponding status is returned.

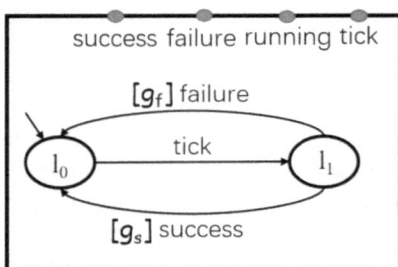 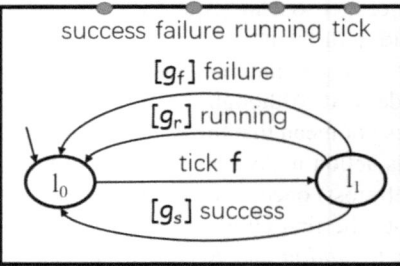

Fig. 2. The BIP components for the condition (left) and action (right) node.

Action Node. Given an action node $\mathcal{N} = \langle \mathcal{V}_d, \mathcal{V}_n, A, g_s, g_f, g_r, f \rangle$, it is transformed into a BIP component $\mathcal{B}_{\mathcal{N}} = \langle \mathcal{V}', \mathcal{L}, \mathcal{P}, \mathcal{T}, \ell \rangle$, where

- $\mathcal{V}' = \mathcal{V}_d$;

- $\mathcal{L} = \{l_0, l_1\}$;
- $\mathcal{P} = \{tick, success, failure, running\}$;
- $\mathcal{T} = \{(l_0, tick, true, f, l_1), (l_1, success, g_s, null, l_0), (l_1, failure, g_f, null, l_0), (l_1, running, g_r, null, l_0)\}$;
- $\ell = l_0$.

The component is shown in the right part of Fig. 2. The transformation for locations and ports are similar to these of the condition node. From the initial location l_0, it can take the transition labelled by port *tick* and performs the computation task represented by function f on receiving a ticking signal. From location l_1 there are three possible outgoing transitions. If the task completes successfully and condition g_s holds, then the transition labelled by port *success* is taken to return a success status. If the task fails and condition g_f holds, then the transition labelled by port *failure* to return a failure status. Otherwise, a running status is returned.

Sequence Node. Given a sequence node $\mathcal{N} = \langle \mathcal{V}_d, \mathcal{V}_n, S, g_s, g_f, g_r, f \rangle$, it is transformed into a BIP component $\mathcal{B}_\mathcal{N} = \langle \mathcal{V}', \mathcal{L}, \mathcal{P}, \mathcal{T}, \ell \rangle$, where

- $\mathcal{V}' = \mathcal{V}_d$;
- $\mathcal{L} = \{l_0, l_f, l_r, l_{|\mathcal{V}_n|+1}\} \cup \{l_i, l'_i | i \in [1, |\mathcal{V}_n|]\}$;
- $\mathcal{P} = \{tick, success, failure, running\} \cup \{tick_i, success_i, failure_i, running_i \mid i \in [1, |\mathcal{V}_n|]\}$;
- $\mathcal{T} = \{(l_0, tick, true, null, l_1), (l_f, failure, true, null, l_0), (l_r, running, true, null, l_0), (l_{|\mathcal{V}_n|+1}, success, true, null, l_0)\} \cup \{(l_i, tick_i, true, null, l'_i), (l'_i, success_i, true, null, l_{i+1}), (l'_1, failure_i, true, null, l_f), (l'_i, running_i, true, null, l_r) \mid i \in [1, |\mathcal{V}_n|]\}$;
- $\ell = l_0$.

The transformation is parameterized with the number of children. Figure 3 shows an example component for a sequence node with two children. In the general case, the component contains four ports for communicating with its parent, as well as four ports for communicating with each child. Thus, in total there are $4 \times |\mathcal{X}| + 4$ ports for each node, where $|\mathcal{X}|$ is the number of its children nodes.

The transformation for the automaton is according to the logic of a sequence node. When receiving a tick from its parent, it forwards the tick signal to its children from left to right by executing the corresponding transition labelled by port $tick_i$ sequentially. When all children return the success status, it will return a success status. If one child returns a failure or running status, it moves to an intermediate state l_f or l_r, stops ticking the subsequent children and returns a failure or running status accordingly.

Fallback Node. Given a fallback node $\mathcal{N} = \langle \mathcal{V}_d, \mathcal{V}_n, F, g_s, g_f, g_r, f \rangle$, it is transformed into a BIP component $\mathcal{B}_\mathcal{N} = \langle \mathcal{V}', \mathcal{L}, \mathcal{P}, \mathcal{T}, \ell \rangle$, where

- $\mathcal{V}' = \mathcal{V}_d$
- $\mathcal{L} = \{l_0, l_f, l_r, l_{|\mathcal{V}_n|+1}\} \cup \{l_i, l'_i | i \in [1, |\mathcal{V}_n|]\}$
- $\mathcal{P} = \{tick, success, failure, running\} \cup \{tick_i, success_i, failure_i, running_i | i \in [1, |\mathcal{V}_n|]\}$

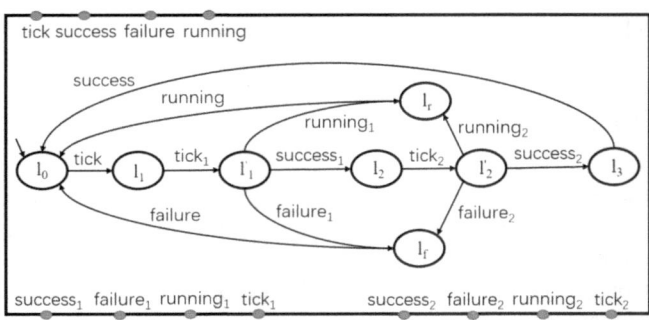

Fig. 3. The BIP component for a sequence node with two children.

- $\mathcal{T} = \{(l_0, tick, true, null, l_1), (l_f, failure, true, null, l_0),$
  $(l_r, running, true, null, l_0), (l_{|\mathcal{V}_n|+1}, success, true, null, l_0)\}$
  $\bigcup\{(l_i, tick_i, true, null, l_i'), (l_i', success_i, true, null, l_{i+1}),$
  $(l_1', failure_i, true, null, l_f), (l_i', running_i, true, null, l_r)|i \in [1, |\mathcal{V}_n|]\},$
- $\ell = l_0.$

The transformation is similar to that of the sequence node. The major difference lies in the transitions for returning the failure and success status. A fallback node returns a success status if one child returns so, while returns a failure status if all children return so. Thus, the automaton of this component can be obtained from the one in Fig. 3 by replacing the transition label *failure* with *success* and vice versa.

Parallel Node. Given a parallel node $\mathcal{N} = \langle \mathcal{V}_d, \mathcal{V}_n, P, g_s, g_f, g_r, f \rangle$, it is transformed into a BIP component $\mathcal{B}_\mathcal{N} = \langle \mathcal{V}', \mathcal{L}, \mathcal{P}, \mathcal{T}, \ell \rangle$, where

- $\mathcal{V}' = \mathcal{V}_d \cup \{num_s, num_f\}$
- $\mathcal{L} = \{l_0, l_{|\mathcal{V}_n|+1}\} \cup \{l_i, l_i'|i \in [1, |\mathcal{V}_n|]\}$
- $\mathcal{P} = \{tick, success, failure, running\} \cup$
  $\{tick_i, success_i, failure_i, running_i|i \in [1, |\mathcal{V}_n|]\}$
- $\mathcal{T} = \{(l_0, tick, true, null, l_1), (l_f, failure, true, null, l_0),$
  $(l_r, running, true, null, l_0), (l_{|\mathcal{V}_n|+1}, success, true, null, l_0)\}$
  $\bigcup\{(l_i, tick_i, true, null, l_i'), (l_i', success_i, true, null, l_{i+1}),$
  $(l_1', failure_i, true, null, l_f), (l_i', running_i, true, null, l_r) \mid i \in [1, |\mathcal{V}_n|]\};$
- $\ell = l_0.$

The transformation is parameterized with the number of children $\mathcal{C}$ and also the success threshold M. As an example, Fig. 4 shows a transformation of a parallel node with two children. Two additional local variables num_s and num_f are introduced in the variable set $\mathcal{V}'$, in order to record the number of succeed and failed children, respectively. The component returns a success status if the number of successful children reaches the threshold M (i.e., $num_s = M$), and returns a failure status if more than $|\mathcal{V}_n| - M$ children fail (i.e., $num_f > |\mathcal{V}_n| - M$). For the other cases, i.e., $\neg(num_s = M) \wedge \neg(num_f > |\mathcal{V}_n| - M)$, it returns a running status.

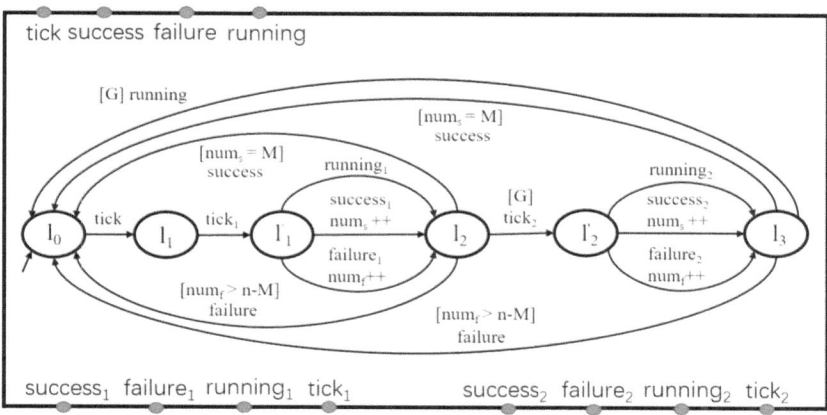

Fig. 4. The BIP component for a parallel node with two children,where $n = |\mathcal{V}_n|$ and $G = (\neg(num_s = M) \wedge \neg(num_f > n - M))$.

5.2 Transformation of BT Edges

Given $\mathcal{BT} = \langle \mathcal{NS}, V, E, L, v_0 \rangle$, the edges of BT are transformed into an interaction model, representing the coordination of the executions of the nodes. Specifically, for each BT edge $e = (v, v') \in E$, we introduce the following four interactions to represent the communication of the ticking signal and the returned status between the BT node $\mathcal{N} = L(v)$ and its child $\mathcal{N}' = L(v')$:

1. $\gamma_{tick}^e = \langle true, \{\mathcal{B}_\mathcal{N}.tick, \mathcal{B}_{\mathcal{N}'}.tick\}, null \rangle$,
2. $\gamma_{success}^e = \langle true, \{\mathcal{B}_\mathcal{N}.success, \mathcal{B}_{\mathcal{N}'}.success\}, null \rangle$,
3. $\gamma_{failure}^e = \langle true, \{\mathcal{B}_\mathcal{N}.failure, \mathcal{B}_{\mathcal{N}'}.failure\}, null \rangle$,
4. $\gamma_{running}^e = \langle true, \{\mathcal{B}_\mathcal{N}.running, \mathcal{B}_{\mathcal{N}'}.running\}, null \rangle$.

The interaction model is then formally defined by the set $\Gamma_{\mathcal{BT}} = \{\gamma_{tick}^e, \gamma_{success}^e, \gamma_{failure}^e, \gamma_{running}^e \mid e \in E\}$. The transformation of the entire BT is obtained by a parallel composition of components using the interaction model, which is defined by a tuple $\mathcal{M}_{\mathcal{BT}} = \langle \{\mathcal{B}_\mathcal{N} \mid \mathcal{N} \in \mathcal{NS}\}, \Gamma_{\mathcal{BT}} \rangle$.

As explained in Sect. 3.1, the execution of a BT is triggered by a ticking signal, which is generated by the unique root node periodically. There are two possible ways to handle this ticking signal in the BIP transformation. The straightforward way is to introduce one more component with a single state, a single port *tick* and a self-loop transition labelled by port *tick*. This port is then synchronized with the tick port of the root component to represent the passing of the ticking signal. Another simple way that can avoid creating new component is to change the tick port of the root component to be an internal transition. The internal transitions in BIP semantics are always enabled from the source states, and there is no need to synchronize with some other triggers. In this work, we take the second way to mimic the passing of ticking signals.

Example 3. Considering again the BT in Fig. 1, the BIP model encoding this BT is illustrated in Fig. 5. For readability, we omit drawing the states and transitions of the components, but only show their interfaces and interactions. The binary interactions (i.e., synchronizations) are denoted by solid lines in blue. It is straightforward to see that after the transformation, the number of components equals to the number of tree nodes and all components of the same type have the same interface. Notice that the t*tick* port of the root sequence node is marked as an internal transition in order to mimic the generation of ticking signals.

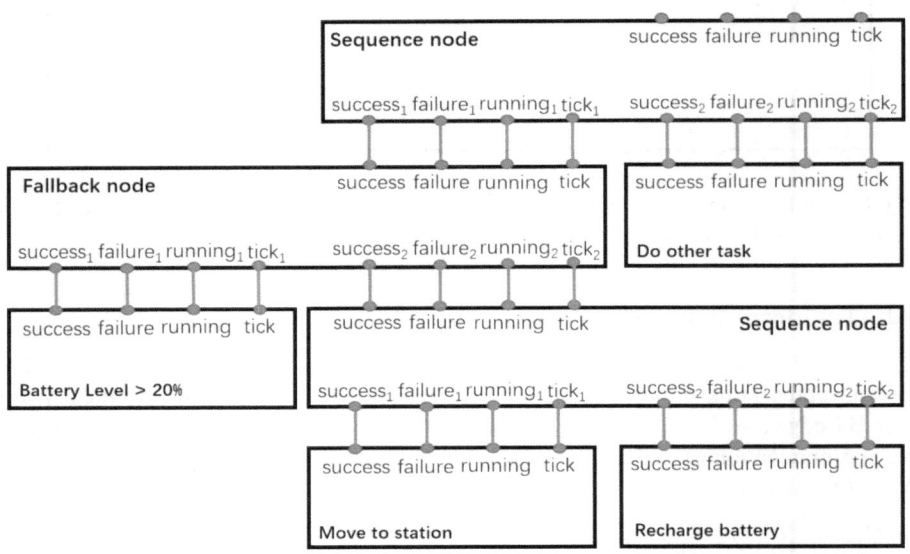

Fig. 5. The BIP model of the BT in Fig. 1.

The complexity of the transformation is linear to the size of the tree. In the general case, given a BT with m nodes and k edges, this transformation produces a BIP model with m components and $4 \times k$ interactions. Moreover, in contrast to the hierarchical encoding presented in [12, 13], all the components in our transformation are organized in a flat manner without introducing any hierarchy. It is also compositional since all components of the same node type have the same interface. One can compose two nodes by connecting their ports using interactions without changing the other components.

5.3   Operational Semantics and Correctness Proof

Given the formal model $\mathcal{M}_{\mathcal{BT}} = \langle \{ \mathcal{B}_\mathcal{N} \mid \mathcal{N} \in \mathcal{NS} \}, \Gamma_{\mathcal{BT}} \rangle$, we define its operational semantics in terms of a labelled transition system. A state of the labelled transition system is denoted by a tuple $c = \langle \langle l_1, \mathbf{V}_1 \rangle, \ldots, \langle l_n, \mathbf{V}_n \rangle \rangle$, where $\forall i \in [1, n], l_i \in \mathbb{L}_i$ and $\mathbf{V}_i$ is an evaluation of local variables $\mathcal{V}_i$ of component $\mathcal{B}_i$. For simplicity, we denote by $\mathcal{C}$ the set of all possible states. A system state c is initial if $\forall i \in [1, n], l_i = \ell_i$ and $\mathbf{V}_i$ is the initial evaluation. An interaction $\gamma = \langle g, \mathcal{P}, f \rangle \in \Gamma$ is enabled on a system state $c = \langle \langle l_1, \mathbf{V}_1 \rangle, \ldots, \langle l_n, \mathbf{V}_n \rangle \rangle$ if (1) $\bigwedge_{i=1}^{n} \mathbf{V}_i \models g$, i.e., guard g is satisfied by the evaluation of the variables; and (2) for each automaton $\mathcal{B}_i = \langle \mathcal{V}_i, \mathcal{L}_i, \mathcal{P}_i, \mathcal{T}_i, \ell_i \rangle, i \in [1, n]$ such that $\mathcal{P} \cap \mathcal{P}_i \neq \emptyset$, there is an enabled transition labeled by $\mathcal{P} \cap \mathcal{P}_i$, i.e., $\langle l_i, \mathcal{P} \cap \mathcal{P}_i, g_i, f_i, l_i' \rangle \in \mathcal{T}_i$ and $\mathbf{V}_i \models g_i$. Notice that it is possible to have multiple interactions enabled on a given state. The execution of the enabled interactions is then chosen in a nondeterministic way.

Definition 5. (Operational semantics). *Given the formal model of a $\mathcal{BT}$ $\mathcal{M}_{\mathcal{BT}} = \langle \{\mathcal{B}_\mathcal{N} \mid \mathcal{N} \in \mathcal{NS}\}, \Gamma_{\mathcal{BT}} \rangle$, the operational semantics is defined by a labelled transition system $\mathcal{TS} = \langle \mathcal{C}, \Sigma, \mathcal{R}, \mathcal{C}_0 \rangle$, where*

1. *$\mathcal{C}$ is a set of states;*
2. *Σ is a set of transition labels and $\Sigma = \Gamma_{\mathcal{BT}}$;*
3. *$\mathcal{R} \subseteq \mathcal{C} \times \Sigma \times \mathcal{C}$ is a set of transitions;*
4. *$\mathcal{C}_0$ is the set of initial states.*

There is a transition from one state $c = \langle \langle l_1, \mathbf{V}_1 \rangle, \ldots, \langle l_n, \mathbf{V}_n \rangle \rangle$ to another state $c' = \langle \langle l_1', \mathbf{V}_1' \rangle, \ldots, \langle l_n', \mathbf{V}_n' \rangle \rangle$, if there is an interaction $\gamma = \langle g, \mathcal{P}, f \rangle \in \Gamma_{\mathcal{BT}}$ such that

1. γ is enabled on state c;
2. for each automaton $\mathcal{B}_i$ such that $\mathcal{P} \cap \mathcal{P}_i \neq \emptyset$, there is an edge $\langle l_i, \mathcal{P} \cap \mathcal{P}_i, g_i, f_i, l_i'' \rangle \in \mathcal{T}_i$, then $l_i' = l_i''$ and $\mathbf{V}_i' = \mathbf{V}_i[\mathbb{V}/f_i(\mathbb{V})]$;
3. for each automaton $\mathcal{B}_i$ such that $\mathcal{P} \cap \mathbb{P}_i = \emptyset$, then $l_i' = l_i$ and $\mathbf{V}_i' = \mathbf{V}_i$.

Given a state c, we denote by $Post(c)$ the set of successor states, i.e., $Post(c) = \{c' \mid \exists \gamma \in \Gamma_{\mathcal{BT}}. (c, \gamma, c') \in \mathcal{R}\}$. A path π is a sequence of states $c_0, c_1, c_2, \ldots$, where the starting state is the initial state, i.e., $c_0 \in \mathcal{C}_0$ and $\forall i \geq 0, (c_i, \gamma, c_{i+1}) \in \mathcal{R}$ for some interaction $\gamma \in \Gamma$. We adopt the notation $\pi(i) = c_i$ to denote the ith state of the sequence, and $\pi(..i) = c_0, c_1, \ldots, c_i$ to denote the ith prefix. A state c_n is reachable if there is a path leading from an initial state to c_n. The set of reachable states is denoted by $\mathcal{C}_R$.

Theorem 1. *Given a $\mathcal{BT}$, let $\mathcal{M}_{\mathcal{BT}}$ be the corresponding BIP model. For every execution of $\mathcal{BT}$, there is an equivalent path of $\mathcal{M}_{\mathcal{BT}}$.*

Proof. The proof is by induction on the structure of BT. Recall that the execution of BT is triggered by the ticking signals periodically. We characterize an execution of BT by a sequence of tuples $(tick_i, status_i)$, where each $tick_i$ is a discrete time and $status_i \in \{success, failure, running\}$. First we consider the tree with only one leaf node. In the case of an action node, and for each $tick_i$ and $status_i$, there is a path l_0, l_1, l_0 in the corresponding BIP component according to the transformation in Fig. 2, where transition from l_0 to l_1 is labelled by port $tick$ and the subsequent transition from l_1 to l_0 is labelled by a port of the returned status $status_i$. The proof applies to the condition node naturally.

Then we consider the case where $\mathcal{BT}$ is composed of a control node and n children nodes. We take the sequence node as an example. The proof applies to fallback and parallel nodes in a similar manner. Let $\mathcal{N}$ be the sequence node connected to n children nodes $\mathcal{N}_1, \ldots, \mathcal{N}_n$ from left to right, and let the corresponding BIP components be $\mathcal{B}, \mathcal{B}_1, ..., \mathcal{B}_n$, respectively. Assume that the result holds for all children nodes, we need to prove that it also holds for the tree rooted in the sequence node. Specifically, for each $(tick_i, status_i)$ in an execution of the sequence node $\mathcal{N}$, we prove that there is a path in the corresponding BIP component $\mathcal{B}$. In case that $status_i = success$, according to the assumption, for each child node $\mathcal{N}_i$ there is a path that executes port $success$ of $\mathcal{B}_i$. Then by composing component $\mathcal{B}$ with $\mathcal{B}_i$, we obtain a model which contains an interaction $\gamma_{success} = \langle true, \{\mathcal{B}.success, \mathcal{B}_i.success\}, null \rangle$ and according to the operational semantics, interaction $\gamma_{success}$ is enabled on a state where component $\mathcal{B}$ stays at location l_{n+1}. Thus, $\mathcal{B}$ can execute port $success$. For the other cases where $status_i \in \{failure, running\}$, the proofs are similar.

6   Formal Verification of BT

This section is concerned with the following problem: given a $\mathcal{BT}$ and a safety property ϕ, how can we verify whether $\mathcal{BT}$ satisfies ϕ automatically?

In order to solve the above problem, we have built a tool-chain as shown in Fig. 6. As the inputs, the tool-chain takes a BT described in the XML format and a safety property specificaiton in LTL. The front-end of the tool-chain implements the proposed model transformation from BT to BIP and the back-end integrates the BIP model checking tool that the authors have developed previously [5]. When checking whether the input BT satisfies the property specification, the tool-chain can produce either a correctness proof when the property specification is satisfied, or a counterexample, otherwise. The counterexample can be further used for the correction of the input BT. We remark that comparing with the work in [3], which can only generate a single abstract state to demonstrate the unsatisfibility of LTL formulae, the proposed tool-chain can produce an execution path as the counterexample, which shows how the BT executes from the initial state to the state violating the property.

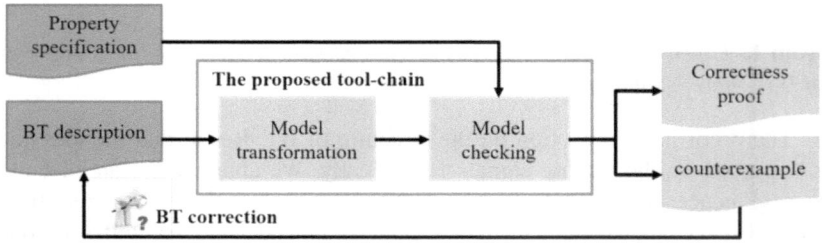

Fig. 6. The proposed tool-chain for automated verification of BT.

We consider two practical BT examples to demonstrate the effectiveness of the proposed method. Due to the space limits, we omit elaborating the formal models of the related BTs and refer to the Github repository[1] for both the source code of the implementation and the models.

---

[1] https://github.com/789wpw/xml2bip.

Example 4. The first example BT shown in Fig. 7 is used to control a Mars-rover-like robot [3]. The robot is equipped with a solar panel and it is unfolded to charge during the day. When a storm comes, the robot needs to fold the panel and enter a hibernating state to protect itself from being damaged. We assume that the robot does not consume power while hibernating.

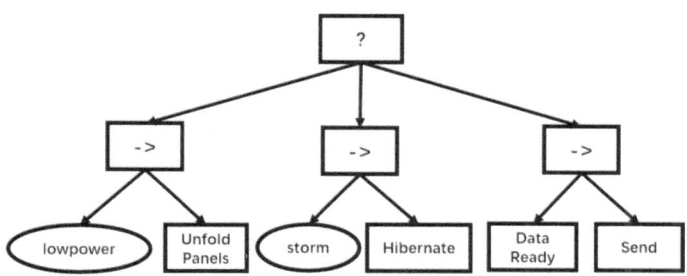

Fig. 7. An example BT controlling the Mars-rover-like robot.

The considered safety property is that the robot should never be damaged. That is, if there is a storm, the robot cannot unfold the panels to charge. First of all, we formally specify this safety property as a LTL formula. Let propositions *storm* and *hibernating* denote the occurrence of a storm and the robot being at hibernating state, respectively. Let G denote the temporal operator "globally" with the usual meaning [8]. Then the property can be formalized as the following formula: $\phi = G(\neg(storm \wedge \neg hibernating))$.

Our tool-chain is able to check that the BT in Fig. 7 does not satisfy this safety property, and it also produces a counterexample as a sequence of states explaining the violation. The scenario is where there is storm initially and the robot chooses to unfold panels without checking the occurrence of storm. A possible correction is to modify the BT by swapping the sub-tree of unfolding panels with that of hibernating, such that the storm check is performed before the power check and the panels are unfolded only if there is no storm.

Example 5. The second example BT shown in Fig. 8 is used for train speed control and door management. Let integer variable v denote the train speed, boolean variable s denote the status of the train being stationary, and boolean variable d denote the status of the door being open, respectively. The formal definitions of leaf nodes are given in Table 2. We assume that the action nodes *stop* and *open* will always complete successfully. We also assume a function f that increases and decreases the speed of the train in the legal interval when the action *move* is running.

The considered safety property is that whenever the train stops and the door is open, the speed must be zero. Formally, this property can be specified as a LTL formula $G(s = true \implies v = 0) \wedge (d = true \implies v = 0)$. By using the proposed tool-chain, we can verify that this safety property is not satisfied by the BT in Fig. 8.

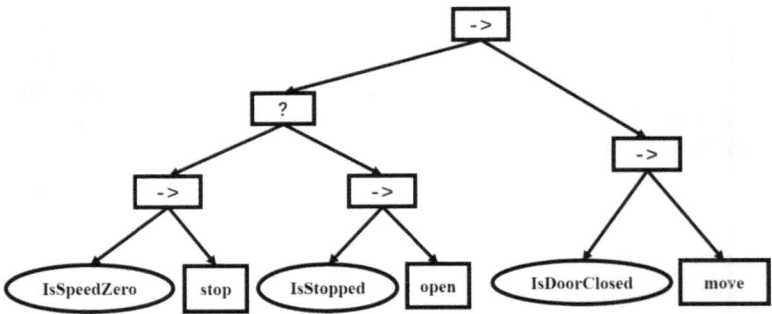

Fig. 8. An example BT for the train speed control and door management.

Table 2. The definition of leaf nodes in Fig. 8

Node	Formal definition
IsSpeedZero	$\langle\{v\}, C, v = 0, v \neq 0, false, null\rangle$
IsStopped	$\langle\{s\}, C, s = true, s = false, false, null\rangle$
IsDoorClosed	$\langle\{d\}, C, d = false, d = true, false, null\rangle$
stop	$\langle\{s\}, \emptyset, A, true, false, false, s := true\rangle$
open	$\langle\{d\}, \emptyset, A, true, false, false, d := true\rangle$
move	$\langle\{v\}, \emptyset, A, false, false, true, v := f(v)\rangle$

Considering the function f increases the speed to a given limit with a constant accelera-tion, a counterexample shows that from a satisfying state $(d = false, v = 0, s = true)$, the train is possible to reach a falsifying state $(d = true, v = 1, s = true)$, where the speed is greater than 0 and the door is open. We remark that this BT cannot be verified by the method presented in [3], due to its inability of encoding the computations in both action and condition nodes.

7   Conclusion and Future Work

This work presents a formal verification method for BT based on a compositional model transformation of BT to BIP. The proposed method enables automated verifi-cation of BT using state-of-the-art model checking techniques without compromising the modularity and flexibility of BT. As the future work, we would like to apply the pro-posed method and tool-chain to a larger set of BT implemented in practical libraries, e.g., BehaviorTree.CPP[2]. And we would also like to explore how to utilize the code generation capability offered by the BIP framework [1,2] such that executable C/C++ codes can be generated automatically for BT.

---

[2] https://github.com/BehaviorTree/BehaviorTree.CPP.

References

1. Abdellatif, T., Bensalem, S., Combaz, J., Silva, L.d., Ingrand, F.: Rigorous design of robot software: a formal component-based approach. Rob. Auton. Syst. **60**(12), 1563–1578 (2012)
2. Basu, A., et al.: Rigorous Component-Based System Design Using the BIP Framework. IEEE, Software (2011)
3. Biggar, O., Zamani, M.: A framework for formal verification of behavior trees with linear temporal logic. IEEE Rob. Autom. Lett. **5**(2), 2341–2348 (2020)
4. Biggar, O., Zamani, M., Shames, I.: An expressiveness hierarchy of behavior trees and related architectures. IEEE Robot. Autom. Lett. **6**(3), 5397–5404 (2021)
5. Bliudze, S., et al.: Formal verification of infinite-state BIP models. In: Finkbeiner, B., Pu, G., Zhang, L. (eds.) ATVA 2015. LNCS, vol. 9364, pp. 326–343. Springer, Cham (2015). https://doi.org/10.1007/978-3-319-24953-7_25
6. Cavada, R., et al.: The NUXMV symbolic model checker. In: Biere, A., Bloem, R. (eds.) CAV 2014. LNCS, vol. 8559, pp. 334–342. Springer, Cham (2014). https://doi.org/10.1007/978-3-319-08867-9_22
7. Chehida, S., Baouya, A., Bensalem, S.: Component-based approach combining UML and BIP for rigorous system design. In: Salaün, G., Wijs, A. (eds.) FACS 2021. LNCS, vol. 13077, pp. 27–43. Springer, Cham (2021). https://doi.org/10.1007/978-3-030-90636-8_2
8. Clarke, E.M., Henzinger, T.A., Veith, H., Bloem, R. (eds.): Handbook of Model Checking. Springer International Publishing, Cham (2018). https://doi.org/10.1007/978-3-319-10575-8
9. Colledanchise, M., Cicala, G., Domenichelli, D.E., Natale, L., Tacchella, A.: Formalizing the execution context of behavior trees for runtime verification of deliberative policies. In: 2021 IEEE/RSJ International Conference on Intelligent Robots and Systems (IROS) (2021)
10. Colledanchise, M., Murray, R.M., Ögren, P.: Synthesis of correct-by-construction behavior trees. In: IEEE/RSJ International Conference on Intelligent Robots and Systems, pp. 6039–6046 (2017)
11. Colledanchise, M., Natale, L.: On the implementation of behavior trees in robotics. IEEE Robot. Autom. Lett. **6**(3), 5929–5936 (2021)
12. Colledanchise, M., Ögren, P.: Behavior trees in robotics and AI: an introduction. CRC Press (2018)
13. Colledanchise, M., Ögren, P.: How behavior trees modularize hybrid control systems and generalize sequential behavior compositions, the subsumption architecture, and decision trees. IEEE Trans. Rob. **33**(2), 372–389 (2017)
14. Colvin, R.J., Hayes, I.J.: A semantics for behavior trees using CSP with specification commands. Sci. Comput. Program. **76**(10), 891–914 (2011)
15. Henn, T., Völker, M., Kowalewski, S., Trinh, M., Petrovic, O., Brecher, C.: Verification of behavior trees using linear constrained horn clauses. In: Groote, J.F., Huisman, M. (eds.) Formal Methods for Industrial Critical Systems: 27th International Conference, FMICS 2022, Warsaw, Poland, September 14–15, 2022, Proceedings, pp. 211–225. Springer International Publishing, Cham (2022). https://doi.org/10.1007/978-3-031-15008-1_14
16. Iovino, M., Scukins, E., Styrud, J., Ögren, P., Smith, C.: A survey of behavior trees in robotics and AI. Robot. Auton. Syst. **154**, 104096 (2022)
17. Klöckner, A.: Interfacing behavior trees with the world using description logic. In: AIAA Guidance, Navigation, and Control Conference, p. 4636 (2013)
18. Konnov, I., Kotek, T., Wang, Q., Veith, H., Bliudze, S., Sifakis, J.: Parameterized systems in BIP: design and model checking. In: Proceedings of the 27th International Conference on Concurrency Theory (2016)

19. Marzinotto, A., Colledanchise, M., Smith, C., Ögren, P.: Towards a unified behavior trees framework for robot control. In: 2014 IEEE International Conference on Robotics and Automation (ICRA), pp. 5420–5427 (2014)
20. Nussbaumer, C.J.M., Kieliger, L.: Bidirectional transformation between BIP and SysML for visualisation and editing (2017). http://infoscience.epfl.ch/record/227431
21. Serbinowska, S.S., Johnson, T.T.: BehaVerify: verifying temporal logic specifications for behavior trees. In: Schlingloff, B.-H., Chai, M. (eds.) Software Engineering and Formal Methods: 20th International Conference, SEFM 2022, Berlin, Germany, September 26–30, 2022, Proceedings, pp. 307–323. Springer International Publishing, Cham (2022). https://doi.org/10.1007/978-3-031-17108-6_19

Verified Fault Handling for Modern Board Management Controllers

Ben Fiedler(✉) , Zikai Liu , David Cock , and Timothy Roscoe

ETH Zürich, Zürich, Switzerland
ben.fiedler@inf.ethz.ch

Abstract. Fault handling is the timely and crash-free response to critical changes in a system's operating characteristics, such as rapid temperature increases, or electrical shorts. In a typical computer system, it is the board management controller's job to correctly respond to such anomalous situations.

We develop an Isabelle/HOL model of a state machine for fault handling and define semantics for correctness of this procedure. Additionally, we formalize a notion of refinement that allows us to prove the correctness of implementations of this state machine. We also provide the first verified implementation of a C-based fault handler for board management controllers. Our implementation and the accompanying proofs are open-sourced and available online.

Furthermore, we successfully deploy our verified fault handler on top of the seL4 microkernel and alongside a production-grade, open source software stack widely deployed today, applying the cyber-retrofit approach to securing board management controllers in practice. The implementation and proof effort required is moderate, and our efforts indicate that already a small team of a handful of people can significantly raise the level of assurance of a modern, highly privileged software system.

Keywords: fault handling · applied formal methods · low-level systems verification · board management controllers

1 Introduction

Board management controllers (BMCs) are the hidden centerpieces of modern computers, and present a highly privileged interface to the lowest-level bits of hardware, such as clock, power, and fan control. Software running on the BMC is responsible for the safe operation of the platform, and often also for remote management. BMCs are critical pieces of hardware that *must* run correctly, otherwise the hardware can be irreversibly damaged.

Despite this, BMCs today are not built to be trustworthy and present a good candidate for the cyber-retrofit approach [24] to building high-assurance systems, by identifying critical functionality that can be extracted from a legacy system and verified or otherwise improved, while leaving the rest of the functionality untouched.

We apply the cyber-retrofit approach to the fault handling procedure of a BMC firmware image based on OpenBMC and Linux. *Fault handling* is the process of interpreting and responding to faults: hardware-generated events corresponding to a change

© The Author(s), under exclusive license to Springer Nature Switzerland AG 2024
D. Marmsoler and M. Sun (Eds.): FACS 2024, LNCS 15189, pp. 21–38, 2024.
https://doi.org/10.1007/978-3-031-71261-6_2

in the external system: over-temperature, over-current, under-voltage, and so on. This is a critical function of the BMC, and it must always run reliably. We verify the fault handling implementation written in C against an abstract specification written in Isabelle/HOL, proving that our implementation is functionally correct.

Our definition of functional correctness states that any received fault will be handled in bounded time and, furthermore, that occurrence of a *critical* fault will eventually lead to a system shutdown. Our proof uses established tools for verification of C programs in Isabelle/HOL, developed for the verification of the seL4 microkernel, but applies them in a new context.

Furthermore, we have deployed our implementation on a real hardware platform: a server-class research computer we describe in Sect. 7. The fault handler runs as a component on the seL4 [23] formally verified microkernel, while the remaining legacy OpenBMC/Linux firmware runs isolated within a virtual machine (VM). Our source code and the associated proofs are available online on Zenodo [16].

2   Background

Most of the integrated circuits (ICs) in a modern server are connected by one or more I^2C buses [35], which provide a simple, two-wire protocol for transferring data between ICs and *bus controllers*. The System Management Bus (SMBus) [33] and Power Management Bus (PMBus) [38] standards define data transfer operations and semantics on top of the I^2C protocol, along with a procedure to report faults to bus controllers. The ICs which are networked with I^2C include the voltage regulators and clock distributors which provide power and clock to the rest of the platform.

The I^2C bus controllers are typically part of the BMC, a small computer which thus controls all aspects of the rest of the machine, including CPU and RAM. BMCs are also often network-facing, since remote management capabilities like console access and reset are indispensable for modern platform management.

A fault event is raised by an IC to signal a (possibly critical) change in operating characteristics, e.g. exceeding specified temperature, current, or voltage limits, and is signaled to the BMC generating an interrupt. The BMC must then identify which component faulted, what occurred, and what action must be taken. Figure 1 shows a simplified example of a server design, representative of the machine in Sect. 7, with I^2C buses and the fault-handling signal path.

The faulting IC may itself take action. An over-voltage fault on a voltage controller, for example, might cause it to disable its output automatically, since I^2C communication between the IC and the BMC is too slow to safely react to these real-time events. Once a fault happens, however, the BMC must take appropriate action, for example increasing fan speed in response to a temperature fault. Ultimately, after a fault the BMC must decide whether to shut down the platform to prevent damage to critical components.

Ensuring BMCs are actually *trustworthy* is thus of critical importance. However, they are not built for high assurance [36]. BMCs are typically provided by the motherboard manufacturer as proprietary, closed-source components, making them difficult to inspect or modify. As a result, new exploits appear each year [9–14]. State-of-the-art BMCs today use open-source software like u-bmc [5] or OpenBMC [1], based on an

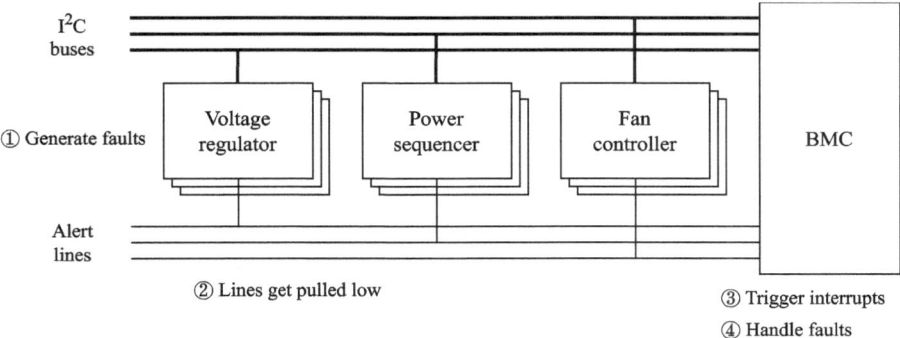

Fig. 1. Overview of the fault handling path on a typical server. I^2C-connected ICs generate faults by pulling the SMBus alert lines connected to the BMC, raising an interrupt.

embedded Linux kernel. These systems provide no rigorous guarantees, and the fact that the software must be customized for each board further reduces reliability and trustworthiness.

We improve on the state of the art by isolating a critical component, namely fault handling, from our system, providing an implementation with provably correct operation, and sandboxing the rest of the legacy system.

3 Related Work

The cyber-retrofit approach [24] to incremental verification of mixed-criticality high-assurance systems was described by the seL4 [23] team in the context of the SMACCM component [7] of the larger DARPA HACMS project. The initial application was the isolation of an unmanned drone's mission-critical real-time flight-control software from a larger, untrusted Linux virtual machine providing lower-criticality functions. The specific approach has since been replicated in other aerial platforms [15], and falls within the wider category of static partitioning approaches [28] to building mixed-criticality systems [4], and has been cited as a reference point for future high-assurance systems for spaceflight [8]. We apply the technique to high-assurance firmware development in response to the shared characteristic of a mixed-criticality composition of safety-critical functions within a larger existing low-assurance code base.

Our proofs cover a subset of the guarantees established as part of the DARPA HACMS project. Compared to the work done by the seL4 team, our verification is limited to one component, thus we are not concerned with properties such as information flow control. We construct specifications for calls to inter-component interfaces, however we do not prove any properties of their implementations.

BMC software stacks have not been widely studied in the literature, and indeed it has been only a few years since open-source BMC software has reached significant adoption [17]. Other approaches for verifying low-level C code bases have been tried before, for example in sensor networks [3] using bounded model checking, or industrial control applications [19] using deductive verification tools. Verification is usually done

```
theorem faults_are_eventually_handled:
  defines tr :: "c_state stream"
  shows "alw step ⟶
            weak_scheduling_fairness ⟶
               alw (pending_fault f ⟶ ev (not (pending_fault f))) tr"

theorem critical_faults_eventually_lead_to_power_down:
  defines f :: c_fault and tr :: "c_state stream"
  assumes "is_critical_fault f"
  shows "alw step ⟶
            weak_scheduling_fairness ⟶
               alw (pending_fault f ⟶ ev powered_down) tr"
```

Listing 1. Top level Isabelle/HOL fault handler correctness theorems.

in full on the annotated source code, in contrast to our incremental approach. The scope of verification also varies greatly, from crash-freedom to functional verification.

4   The Top-Level Result

We prove three core results of the overall fault-handling implementation. First, to discharge the well-formedness assumptions of the StrictC dialect we need to show that the C implementation is crash-free, terminates, and has the same core memory-safety and well-defined-behavior properties as the seL4 kernel itself. We then extend this with the two application-specific properties in Listing 1.

These express that first, any fault that is signaled will eventually be handled and that second, any critical fault will eventually cause the system to be shut down. These are expressed as LTL (linear temporal logic) [32] formulae over the top-level state machine model of the system. The syntax is that of the existing LTL formalization available in Isabelle's HOL Library [30]. They correspond to the following formulae in standard LTL:

$$wsf \implies \qquad\qquad \Box(pending\_fault\ f \implies \Diamond(\neg pending\_fault\ f))$$
$$wsf \implies is\_critical\_fault\ f \implies \quad \Box(pending\_fault\ f \implies \Diamond powered\_down)$$

"Assuming that weak scheduling fairness holds, it is always the case that if fault f is pending now, eventually it is not (and has thus been handled), and that if the fault is critical the system is eventually in a powered-down state."

Both statements make an explicit *weak fairness* assumption *wsf*. Lamport's definition of weak fairness for some transition T states that

$$weak\_fairness(T) \equiv \Box \Diamond \neg enabled(T) \lor \Box \Diamond step(T)$$

The transitions of the fault handling state machine are always enabled, and thus the weak fairness assumption simplifies to:

$$wsf \equiv \Box \Diamond consume\_faults \land \Box \Diamond check\_shutdown$$

```
definition enable_source where "enable_source s t = ..."
definition disable_source where "disable_source s t = ..."
definition receive_faults where "receive_faults s t = ..."

definition consume_faults :: "('s, 'f) state rel" where
  "consume_faults s t ≡ t = s(|
      faults := [],
      shutdown_triggered := shutdown_triggered s
          ∨ (list_ex is_critical_fault (faults s))
  |)"

definition check_shutdown where "check_shutdown = ..."

definition step :: "('s, 'f) state rel" where
  "step s t ≡
     s = t ∨ enable_source s t ∨ disable_source s t
          ∨ receive_faults s t ∨ consume_faults s t
          ∨ check_shutdown s t"
```

Listing 2. The abstract state machine.

"At any point in time, both the fault and shutdown handlers will be called (strictly, be the current task) in a finite number of steps."

This liveness assumption is carried explicitly in both top-level correctness statements and expresses a requirement on the scheduler configuration in any deployment. This may be discharged by, for example, employing the strict real-time MCS scheduler [27] now undergoing verification.

By showing that the C implementation both refines the abstract state-machine model and separately preserves these liveness properties, we establish that they hold for the final deployed fault handler. We integrate this fault handler with the production-grade OpenBMC [1] software stack, a Linux distribution widely used for commercial BMCs. The fault handler comprises a set of native seL4 tasks running alongside the rest of the OpenBMC stack which is isolated within a virtual machine.

This is an application of the cyber-retrofit approach [24]. This technique increases the assurance of a software system by extracting a small, trusted part, comprehensively verifying it, and recombining it with the unmodified base system without compromising either the new formal guarantees or the functionality of the unverified component. Despite the presence of untrusted and unverified code in OpenBMC, we can appeal to the verified security properties of the seL4 microkernel to establish that these verified fault-handling guarantees hold even if the network-accessible OpenBMC component is completely compromised.

```
definition weak_scheduling_fairness :: "('s, 'f) state trace ⇒ bool" where
  "weak_scheduling_fairness ≡ alw (ev consume_faults) ∧
                              alw (ev check_shutdown)"
```

Listing 3. Isabelle/HOL definition of the weak scheduling fairness assumption; it is a straightforward translation of the previous LTL formula defining *wsf*.

```
record ('s, 'f) state =
  enabled_sources :: "'s set"    faults :: "'f list"
  shutdown_triggered :: "bool"   power_up :: "bool"

locale fault_handling_assumptions = fixes
  is_critical_fault :: "'f ⇒ bool" and
  max_faults        :: "nat"
```

Listing 4. Fault handler state and ancillary assumptions.

5   The Abstract Specification

Our top-level model is the nondeterministic state machine defined in Listing 2 with next-step relation step. This expresses the nondeterministic composition of steps of the handler (enable_source, disable_source, consume_faults, check_shutdown), steps of the environment (receive_faults), and idle transitions (s = t). The definition of consume_faults is expanded to illustrate the abstract implementation of a component. Here the list of active faults in the prior state is cleared (consumed), and the shutdown flag is set if at least one of those faults was critical or it was already set.

While the model defines a set of possible next states (the image of the relation), any trace of the implementation is a definite sequence of states lying pairwise in the step relation.

Which trace actually executes is a property of the OS scheduler and any other nondeterminism refined by the final implementation. The OS scheduler and environmental nondeterminism determine the order in which steps are executed to form the final trace. This is abstracted in the model and the only property of the trace we rely on is expressed in the Isabelle/HOL formulation of the weak scheduling fairness assumption in Listing 3.

The abstract state over which the machine operates is defined in Listing 4 . The types 's and 'f are parameters and represent a set of *fault sources* and *fault types*, respectively. Fault sources are disabled, in which case the associated fault values will not be signaled, if they are not in the set of enabled_sources. The list of faults records all as-yet-unhandled faults in order of occurrence. The flag shutdown_triggered records whether a shutdown has been initiated (but not necessarily completed), and power_up whether the system is currently powered. Once the shutdown completes, power_up becomes false.

A *locale* is an Isabelle mechanism to express a named bundle of definitions and assumptions [22]. The fault_handling_assumptions locale assumes the existence

```
definition refines ::
  "('c ⇒ 'a) ⇒ 'a rel ⇒ 'c rel ⇒ bool" where
  "refines lift an cn ≡
    (∀s t. cn s t ⟶ an (lift s) (lift t))"
```

Listing 5. Refinement with respect to a lifting function

```
definition sufficiently_live ::
  "('c ⇒ 'a) ⇒ ('c stream ⇒ bool) ⇒ bool"
where
  "sufficiently_live lift LP ≡
    ∀tr. LP tr ⟶ weak_scheduling_fairness (smap lift tr)"

definition liveness_prop ::
  "(('s,'f) model stream ⇒ bool) ⇒ bool"
where
  "∀tr. (alw step' ⟶ weak_scheduling_fairness ⟶ P) tr"

lemma transfer_liveness_prop:
  assumes
    "refines lift step cn"
    "sufficiently_live lift LP"
    "liveness_prop P"
  shows
    "(alw (lift cn) ⟶ LP ⟶ P o (smap lift)) tr"
```

Listing 6. Liveness preservation across refinement

of an otherwise-undefined predicate classifying some faults as critical (requiring shutdown), and that there exists some bound (`max_faults`) on the number of faults that can occur before the handler executes.

We show that the C implementation of the fault handler, as imported into Isabelle using the StrictC tool chain, is a *refinement* [26] of this abstract state machine. This relation is defined precisely in Listing 5 . The *concrete* system cn is a refinement of the *abstract* system an if, given a lifting function from concrete to abstract states, wherever a concrete transition between two states exists, an abstract transition exists between the corresponding lifted states. Informally, this expresses that "Every behavior of the (concrete) implementation is permitted by the (abstract) specification.".

Refinement does not generally preserve liveness. An implementation which does nothing at all trivially refines any specification. More generally it only requires that the steps of the implementation are some subset of those of the specification, but not that any particular step actually happens. We thus extend our definition as shown in Listing 6 to express that a class of liveness properties (those that depend on weak scheduling fairness) are in fact preserved by our refinement.

A *sufficiently-live* property on the implementation is one which if it holds for a concrete trace, weak scheduling fairness holds for the corresponding lifted abstract trace. A *liveness property* (in this context) is one which holds in any (abstract) state where weak scheduling fairness holds.

The *transfer lemma* establishes that given a sufficiently-live property for the concrete system, any liveness property of the abstract system also holds for the concrete when interpreted through the lifting function. The property may be *transferred* from the abstract states to their pre-images under the lifting function. Both of the top-level formulae of Listing 1 satisfy this definition of liveness, and are thus preserved by the refinement.

6   The Proof

Having defined an abstract model for fault handling, we next implement a fault handler in C as a component on top of the seL4 microkernel [23] and its component framework CAmkES [25]. We provide a behavioral proof of each of the fault handler functions in the form of Hoare triples, as well as total correctness (i.e. termination) proofs for invariant preservation. A brief schematic of the whole code-to-proof pipeline from the C source to the top-level result is shown in Fig. 2.

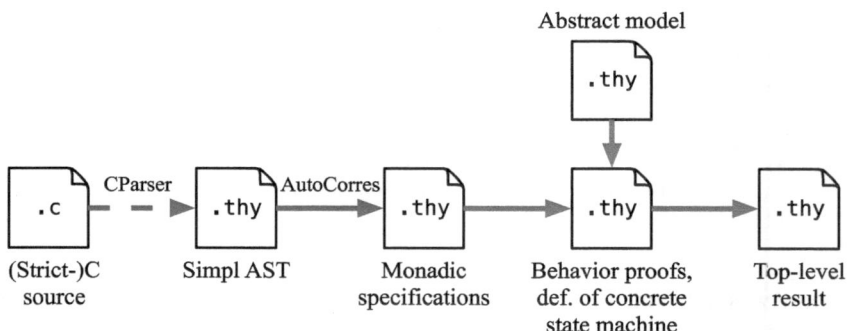

Fig. 2. Overview of the proof pipeline, from C code to final result. The dotted arrow denotes an unverified step, while the solid arrows are formally proven in Isabelle/HOL. Blue and orange distinguish automatic and manual proofs, respectively. (Color figure online)

Our code is written in StrictC, a dialect of the C language developed by Schirmer [34]. As the name suggests, StrictC disallows use of some C constructs, including but not limited to: `goto` statements, assignments within expressions, fall-through cases, type unions, and C99's `bool`. We translate the StrictC program to the Simpl language using an existing, ML-based parser, initially developed for use in the verification of the seL4 microkernel. Simpl is a sequential, imperative programming language with formal semantics defined in Isabelle/HOL. Note that this translation is unverified; there is no formal proof that the semantics of the generated Simpl program matches the semantics of the StrictC program.

The Simpl translation of our C code is then fed into AutoCorres, which was developed by Greenaway et al. [18]. AutoCorres translates the Simpl programs to a monadic

```
lemma inv_no_fail:     "{inv} m {inv}!"
lemma executable:      "executable_specification m"
lemma refines_model:   "{s} m {model.step (lift_state s) (lift_state s')}"
```

Listing 7. Proof obligations for an AutoCorres-generated monadic specification m

specification over a non-deterministic state monad with failure, including an automatically derived proof that these specifications refined the Simpl programs they were generated from.

The monadic specifications generated by AutoCorres contain statements that relate to the well-formedness of the underlying StrictC program: accessed pointers are always valid, satisfy bounds checks when indexing arrays, and so on. Otherwise, the program fails, and the resulting state will have a failure flag set. Showing that the monadic programs do not fail is a significant part of our work. On the other hand, a monadic specification could also be *trivial*, returning an empty set of resulting states. This would imply that said specification cannot be executed. Fortunately, we can automatically prove non-triviality for almost all of our specifications, and complete the remaining cases by hand.

Our main proof effort pertains to proving Hoare triples about the behavior of these monadic specifications. From these behavioral predicates, we construct a state machine, where the state chosen corresponds to the state of the monadic translation of StrictC program. We prove that each monadic specification, translated from its original function, refines a corresponding transition in the abstract state machine. Finally, we assume the existence of a weakly fair scheduler for some transitions, in order to meet the liveness assumptions made about our state machine.

Plugging our refinement results, proven on the abstract machine, together with the concrete behavior proved on our monadic translations, allows us to prove that the concrete machine satisfies the conditions for fault handling identified in Sect. 4. Adding the refinement result generated by AutoCorres relates the concrete state machine's behavior down to the StrictC code.

6.1   Reasoning About the Generated Monadic Specifications

Recall that the monadic specifications generated by AutoCorres are non-deterministic and allow failure, thus we have to prove that for each specification, it never fails *and* always produces at least one resulting state. Thus, after translating a C function f to a monadic specification m, we need to prove three things of m: it terminates and does not fail, it is executable, and it refines the behavior of an abstract transition. We prove each of these statements individually, later combining them to a full correctness statement. A brief overview of the proof goals can be seen in Listing 7 .

In order to prove that the specification does not fail, we need to show that any pointers and arrays accessed/indexed by m are valid. To this end, we define a global invariant inv that is preserved by every monadic program, and it does not cause the program to fail. Listing 8 shows an excerpt of our invariant definition.

Proving that the pointer invariants are preserved is straightforward, since our pointers are valid initially, and are never modified. Proving that the number of faults received

```
definition "valid_num_faults g ≡ num_faults g ≤ max_faults"
definition "valid_ptrs g ≡
              is_valid_w8 addr_ptr ∧
              is_valid_w16 status_word_ptr ∧
              (* ... *)"
definition "inv s ≡ valid_ptrs s ∧ valid_num_faults s"
definition "init s ≡ valid_ptrs s ∧ num_faults s = 0 ∧ ..."
lemma init_impl_invariant: "init s ==> inv s" (* ... *)
```

Listing 8. The main fault handler invariant for the concrete implementation.

stays valid is slightly more involved, since we use num_faults as an index when modifying the max_faults-sized buffer of faults. We utilize lemmas for reasoning about word-sized arithmetic in these cases. Most of the proofs can be discharged by the wp tool.

Next, in order to prove that a monadic specification is non-trivial, we have to prove that the set of result states it produces is non-empty. However, doing so would require us to duplicate all proofs we have already done about total correctness of the invariant. Therefore, we prove an equivalent statement, which allows us to re-use our invariant proofs: we prove that each monadic specification also fails when it produces an empty set of output states. Since we have already proven non-failure under the invariant, this is sufficient to prove executability as well. This property can be derived automatically in almost all cases.

Finally, to prove that our specification refines our abstract model, we define a lifting function, which lifts our concrete state to the abstract model state and is shown in Listing 9. The lift_state function is responsible for translating the data representation from C arrays and integers to the richer type system of our state machine using richer types such as lists and sets. Concretely, the faults list of our abstracted state is represented as two variables in our C code: a statically-sized array of faults, and an integer recording the number of faults currently buffered. The set of enabled sources is related to the non-zero entries of underlying C array. The boolean field holding whether a shutdown was triggered is represented as an integer in the underlying C code[1], with zero corresponding to *false*, and any non-zero value corresponding to *true*. Finally, the abstract machine is considered "powered up" when the main power supply is switched on. Other powered components, such as the passive power rail that powers the BMC itself are irrelevant.

To complete our refinement proof, we need to prove that each monadic specification modifies the state in a way consistent with the abstract model. These proofs boil down to proving that some fields have been modified in a specific way, while leaving most of the other fields untouched. Take the receive_fault transition, for example: we prove that most of the global state remains unchanged, while the number of received faults is increased, and new faults are correctly buffered, keeping the old buffer intact. Thus, we

---

[1] This is due to the StrictC parser being unable to parse the stdbool.h header.

```
definition
  lift_state :: "c_state ⇒ (fault_line, fault) Model.state"
where "lift_state g = (|
    enabled_sources    = {s. faults_enabled g.[unat s] ≠ 0},
    faults             = let n = unat (num_faults g)
                            in take n (list_array (faults g),
    shutdown_triggered = shutdown_triggered g ≠ 0,
    power_up           = psu_up (power_state g)
|)"
```

Listing 9. Definition of our lifting function from c_state to abstract model state

```
(* Framing conditions *)
lemma receive_fault_lift_enabled:
  "{%s. P (lift_enabled s)} receive_fault {%s. P (lift_enabled s)}"
lemma receive_fault_lift_shutdown:
  "{%s. P (lift_shutdown s)} receive_fault {%s. P (lift_shutdown s)}"
lemma receive_fault_lift_power_up:
  "{%s. P (lift_power_up s)} receive_fault {%s. P (lift_power_up s)}"

(* Precise behavior *)
lemma receive_fault_lift_faults_prefix:
  "{%s. prefix a (lift_faults s)}
     receive_fault
   {%s. prefix a (lift_faults s)}"
lemma receive_fault_line_disabled:
  "{%s. P (lift_faults s) ∧ 1 ∉ lift_enabled s}
     receive_fault
   {%s. P (lift_faults s))}"

lemma receive_fault_combined:
  "{%s'. s = s'}
     receive_fault
   {%s'. model.step (lift_state s) (lift_state s')}"
```

Listing 10. Proof obligations split into framing and non-framing conditions.

can decompose our proofs about the whole state by considering each field separately, leveraging wp's proof automation to combine the sub-proofs for us.

An example of the individual proof goals for a function receiving faults is shown in Listing 10. Note that in a system with multiple fault sources, there might multiple functions that receive faults, and each of them have to be proven correct separately. Generally, we divide the proofs in two parts, depending on whether we aim to prove a *framing condition* about a particular field, or whether we prove something about the correct modification of some state.

Many of our proof obligations are, in fact, such framing conditions. Those can often be proven automatically using some existing tooling that "crunches" through these (often repetitive) proofs. It was even able to deal with simple loops, which alleviated a non-trivial amount of proof effort. In more complicated cases, however, we had to write proofs by hand.

The behavioral proofs seldom worked out-of-the-box. In order to get those proofs down to a combination of applying the wp tactic and appropriate simplifier invocations, we introduce and prove other Hoare triples or simplification rules. Sometimes, the form of required rules could be straightforwardly derived from the proof state, indicating where the weakest-precondition tactic was "stuck". However, in other cases the proof state was unhelpful in determining a statement that could be useful to the automated tactics, and required us to carefully construct a proof by hand.

As already mentioned, we rely heavily on the weakest precondition tactic wp, as distributed with the nondeterministic state monad, for verification condition generation. In practice, we found it often not necessary to find the weakest precondition, only a sufficiently weak one. Supplying the tool with theorems that are not in weakest precondition form still guarantees soundness, but completeness is lost: it might be the case that some valid goals are not provable, and in those cases we had to re-visit some intermediate proofs we had done. We used this fact in some of our proofs: by supplying non-weakest preconditions, we were able to reduce the precondition's complexity and simplify the remaining proof. This was especially useful in statements that involved branching.

Finally, we construct our state machine by defining a suitable initial state predicate and transition relation, and prove that, together with the lifting function defined earlier, this concrete state machine is a refinement of the abstract state machine introduced in Sect. 5, which concludes our proof.

From here on, state the liveness assumptions on the concrete state machine like we did on the abstract state machine, and prove that it is indeed sufficient to transfer the properties proved on our abstract state machine. This is shown in Listing 11 . Together with the previously proven facts about the abstract, we achieve the main result shown in Sect. 4.

7   Running on Real Hardware

Enzian [6] is a server-class research computer built at ETH Zurich with two main chips, a CPU and a large FPGA as a second NUMA node. The BMC is a Zynq UltraScale+ MPSoC [2], initially running Linaro Linux and a custom distribution of OpenBMC [1] for board management.

The relevant system architecture is shown in Fig. 3. Enzian uses multiple PMBuses equipped with alert signals to communicate with the peripheral devices. We focus on two interesting voltage regulator ICs, the IR3581 [20] and the MAX15301 [29], in this presentation. Applying the cyber-retrofit approach, we encapsulate the Linux-based OpenBMC in a VM over seL4 (with device pass-through) and implement the fault handling code in a native seL4 process using the CAmkES component framework.

The code requires device drivers to communicate with the regulators. Unlike Linux, limited drivers are available on seL4. For ease of implementation, we use Xilinx driver

```
definition step :: "(c_state, unit) nondet_monad" where
 "step ≡ skip <|> receive_fault <|> enable_faults <|> disable_faults
             (* other fault lines omitted *)
             <|> consume_faults <|> condition lift_shutdown shutdown skip"

theorem step_refines_model:
 "model.refines lift_state model.step (run_monad step)"

definition weak_liveness :: "c_state trace ⇒ bool" where
 "weak_liveness ≡ alw (ev (run_monad consume_faults')) aand
              alw (ev (run_monad check_shutdown'))"

lemma sufficiently_live_assumption: "sufficiently_live weak_liveness"
```

Listing 11. The main refinement theorem of our abstracted code.

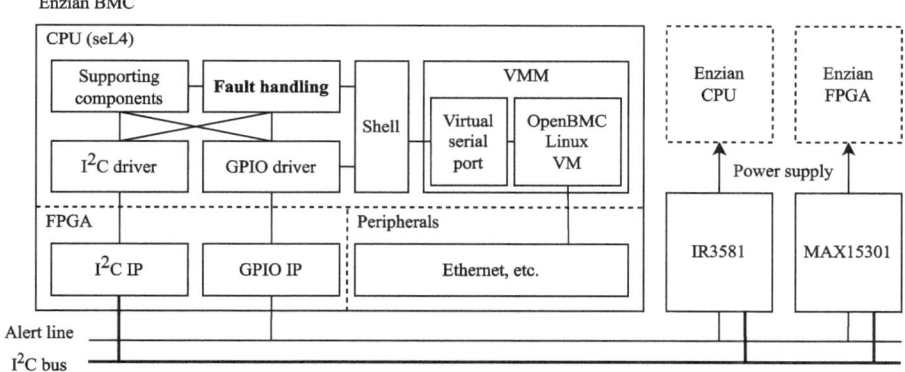

Fig. 3. System architecture.

IPs on the MPSoC FPGA [2] and ported the Xilinx bare-metal drivers [40] to seL4. Currently these components are trusted, and verifying their correctness is beyond the scope of the present work. The fault handler runs in the background; we also implemented an interactive shell for testing. We modify the original OpenBMC to redirect commands to those two regulators to a virtual serial port connected to the native seL4 system. The supporting OpenBMC components include a time server and a board resetting server.

The IR3581 and MAX15301 raise alerts when temperatures exceed set values. To simulate over-temperature alerts reliably, we set the threshold temperatures to *below* room temperature. These alerts are critical faults that should lead to system shutdown.

The evaluation process thus consists of the following steps, performed through the interactive shell:

1. Power up the system.
2. Read back voltage levels to ensure the system is indeed powered up.

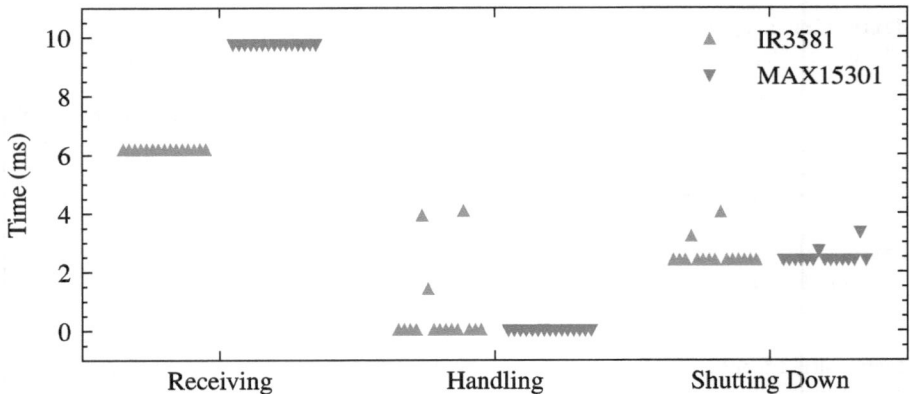

Fig. 4. Time spent in steps of fault handling.

3. Trigger alerts by lowering the threshold temperature. The fault handling code reports time spent in each step to the shell.
4. Read back voltage levels to check if the system is powered down.

In step 3, the SMBus protocol [33] requires the BMC to send out a response-address request on the I^2C bus to identify the faulting device. The device then responds with its address. This step, when receiving faults, involves relatively slow I^2C communication. The same applies to the step of shutting down the system. Time spent during I^2C communication is not a focus of this work. Nevertheless, we show the time of those two steps to briefly demonstrate the end-to-end functionality of the code, while focusing on the step of processing fault events.

For each device, we repeat the procedure 15 times. Lowering the threshold temperatures results in alerts raised by the IR3581 [20] and MAX15301 [29], respectively. Read-back voltages are machine-checked, showing that the fault handling *functions as expected*. The time spent in each fault handling step is reported in Fig. 4.

The fault handling time (middle) is relatively small. Handling one alert from the MAX15301 takes only 25.6 μs on average with a standard deviation of 0.7 μs. For the IR3581, most of the time, handling two alerts takes less than 26.5 μs. However, we observe a few outliers that take up to 4.1 μs. We believe this is due to seL4 switching context between the two alerts. The fault handling and shutting down processes are connected by seL4 signals [39]. When the first alert gets handled, a signal is emitted, which may trigger the seL4 scheduler to deschedule the fault handling code.

Nevertheless, the fault handling code functions correctly on the evaluation platform and in a timely fashion. The outliers for the IR3581 demonstrate how the OS scheduler may affect the liveness of fault handling, motivating a verified mixed-criticality scheduler or careful design in the implementation to better match seL4's default fixed-priority scheduler.

8   Experience

We now briefly discuss our experience replacing the OpenBMC fault handler with our verified implementation.

The entire project, from developing the initial implementation, specification, integrating our code with virtualized OpenBMC, and proving the implementation correct took us about 20 person-months (shared across several people). A detailed breakdown of the individual tasks can be seen in Table 1.

Table 1. Overview of time and lines of proof required for the project. We do not include the existing seL4 components we use.

Task	Effort	Verified Component	LoP	%
Initial seL4 port	6 mo	Top-level statement	167	7.0
verified C	2 mo	Abstract model	354	14.8
Abstract model	1 mo	Extracting monadicspecification from C	43	1.8
Full verification	7 mo	**Proofs about monadicspecifications**	1'733	72.4
Virtualizing OpenBMC and integration	4 mo	Miscellaneous	96	4.0
Total time spent	20 mo	**Total**	2'393	100.0

In total, our efforts clock in at about 7.4k lines of C code (excluding generated files), and 2.4k lines of Isabelle/HOL. Of those 7.5k lines of C, the verification covers about 1k lines, of which 700 lines are source code and 300 lines are headers. Unsurprisingly, the majority of the effort is spent verifying the C code, both in time spent and the proof length.

Verification strongly influenced the implementation during development, for a variety of reasons. On the one hand, writing code that we later verify forces us to be precise with respect to our intended semantics, such as the consideration of what happens when the fault buffer is full. On the other hand, the restrictions placed on us by StrictC disallows some constructs that could come in handy at certain points when interacting with hardware, such as type unions. Finally, it causes us to use language features that alter control flow (such as `continue` or `break` within loops) only very sparingly. It is possible to reason about the specifications for such code, however we found it is often easier to restructure the control flow.

Our proofs themselves are designed to be modular. Changes to the implementation of one function generally have no effect on other functions, assuming we can re-use the same top-level specification. We took care to design our proofs such that some assumptions can easily be changed, such as the definition of a critical fault or the maximum number of buffered faults.

Now that we have verified one component of a larger system, the question arises how to compose our proof with other correctness proofs that we might be interested in developing.

Due to limitations in the StrictC tool, we were forced to specify some C functions, such as those that interact directly with the I^2C hardware through registers. Since we have already come up with these specifications, further proof efforts should be able to leverage these.

However, composing our proofs with correctness proofs of other components that would concurrently execute within the same system is much more difficult, as there is no general method known to compose two unrelated, concurrent specifications.

9 Conclusion

We have shown that the cyber-retrofit strategy can be applied to new domains in order to verify large-scale system. We were able to apply existing verification tooling and techniques to develop and prove correct the fault handling implementation of a modern BMC, while deploying the resulting artifact in production and successfully handling faults.

The required work for writing and proving correct code is sufficiently small to allow even small teams of students to verify non-trivial functionality, thanks to the verification tools developed by the seL4 team [23].

There are open questions for scaling the work in order to verify larger systems. Cross-CAmkES component verification was shown to be possible, and new approaches [31] are being tested to verify sequential code across process boundaries. It is, however, unclear how concurrency fits into the picture. Approaches for verifying concurrent code are being tried right now [21,37], and it remains to be seen how the seL4 team itself will tackle these. We assume a sequential schedule for now.

Future work on our end includes extracting and verifying other parts of the OpenBMC stack, and ensuring the specifications we developed for inter-component function calls. Additionally, we are looking forward to a version of seL4 with a verified mixed-criticality scheduler in order to formalize our liveness assumptions.

Acknowledgments. We thank our anonymous reviewers for their valuable feedback, and helping us to improve the presentation of this paper.

Disclosure of Interests. The authors have no competing interests to declare that are relevant to the content of this article.

References

1. OpenBMC. https://www.openbmc.org/
2. AMD: Zynq UltraScale+ Device Technical Reference Manual. Tech. Rep. 2.4 (2023). https://docs.amd.com/r/en-US/ug1085-zynq-ultrascale-trm
3. Bucur, D., Kwiatkowska, M.: On software verification for sensor nodes. J. Syst. Softw. **84**(10), 1693–1707 (2011)
4. Burns, A., Davis, R.I.: A survey of research into mixed criticality systems. ACM Comput. Surv. **50**(6) (2017). https://doi.org/10.1145/3131347
5. Christian Svensson et al: Open-source firmware for your baseboard management controller (BMC) (2018). https://github.com/u-root/u-bmc, computer Software

6. Cock, D., et al.: Enzian: an open, general, CPU/FPGA platform for systems software research. In: Proceedings of the 27th ACM International Conference on Architectural Support for Programming Languages and Operating Systems, pp. 434–451. ASPLOS '22, Association for Computing Machinery, New York, NY, USA (Feb 2022). https://doi.org/10.1145/3503222.3507742
7. Cofer, D., et al.: Secure mathematically-assured composition of control models. Tech. Rep. AFRL-RI-RS-TR-2017-176, Rockwell Collins, Cedar Rapids, United States (Sep 2017). https://apps.dtic.mil/sti/citations/AD1039782
8. Curbo, J., Falco, G.: A research agenda for space flight software security. In: 2023 IEEE 9th International Conference on Space Mission Challenges for Information Technology (SMC-IT), pp. 68–77 (2023). https://doi.org/10.1109/SMC-IT56444.2023.00016
9. Database, M.C.: CVE-2022-40242 (Dec 2022). https://www.cve.org/CVERecord?id=CVE-2022-40242
10. Database, M.C.: CVE-2022-40259 (Dec 2022). https://www.cve.org/CVERecord?id=CVE-2022-40259
11. Database, M.C.: CVE-2021-39295 (Apr 2023). https://www.cve.org/CVERecord?id=CVE-2021-39295
12. Database, M.C.: CVE-2022-26872 (Jan 2023). https://www.cve.org/CVERecord?id=CVE-2022-26872
13. Database, M.C.: CVE-2023-31008 (Jan 2024). https://www.cve.org/CVERecord?id=CVE-2023-31008
14. Database, M.C.: CVE-2023-31037 (Jan 2024). https://www.cve.org/CVERecord?id=CVE-2023-31037
15. Farrukh, A., West, R.: Flyos: integrated modular avionics for autonomous multicopters. In: 2022 IEEE 28th Real-Time and Embedded Technology and Applications Symposium (RTAS), pp. 68–81 (2022). https://doi.org/10.1109/RTAS54340.2022.00014
16. Fiedler, B., Liu, Z.: An sel4-based board management controller with verified fault handling for the Enzian research computer. https://doi.org/10.5281/zenodo.12775411
17. Frazelle, J.: Opening up the baseboard management controller. Commun. ACM **63**(2), 38–40 (2020)
18. Greenaway, D., Lim, J., Andronick, J., Klein, G.: Don't sweat the small stuff: formal verification of C code without the pain. SIGPLAN Not. **49**(6), 429–439 (2014). https://doi.org/10.1145/2666356.2594296
19. Huisman, M., Monti, R.E.: On the industrial application of critical software verification with VerCors. In: Margaria, T., Steffen, B. (eds.) Leveraging Applications of Formal Methods, Verification and Validation: Applications: 9th International Symposium on Leveraging Applications of Formal Methods, ISoLA 2020, Rhodes, Greece, October 20–30, 2020, Proceedings, Part III, pp. 273–292. Springer International Publishing, Cham (2020). https://doi.org/10.1007/978-3-030-61467-6_18
20. International Rectifier: Dual Output Digital Multi-Phase Controller IR3581. Tech. rep. (Apr 2014)
21. Jung, R., et al.: Iris: Monoids and invariants as an orthogonal basis for concurrent reasoning. In: Proceedings of the 42nd Annual ACM SIGPLAN-SIGACT Symposium on Principles of Programming Languages, pp. 637–650. POPL '15, Association for Computing Machinery, New York, NY, USA (2015). https://doi.org/10.1145/2676726.2676980
22. Kammüller, F., Wenzel, M., Paulson, L.C.: Locales a sectioning concept for isabelle. In: Bertot, Y., Dowek, G., Théry, L., Hirschowitz, A., Paulin, C. (eds.) Theorem Proving in Higher Order Logics, pp. 149–165. Springer Berlin Heidelberg, Berlin, Heidelberg (1999). https://doi.org/10.1007/3-540-48256-3_11
23. Klein, G., et al.: Comprehensive formal verification of an OS microkernel. ACM Transactions on Computer Systems **32**(1), 2:1–2:70 (2014). https://doi.org/10.1145/2560537

24. Klein, G., Andronick, J., Kuz, I., Murray, T., Heiser, G., Fernandez, M.: Formally verified software in the real world. Commun. ACM **61**, 68–77 (2018). https://doi.org/10.1145/3230627
25. Kuz, I., Liu, Y., Gorton, I., Heiser, G.: CAmkES: a component model for secure microkernel-based embedded systems. J. Syst. Softw. **80**(5), 687–699 (2007). https://doi.org/10.1016/j.jss.2006.08.039
26. Lamport, L.: Specifying concurrent program modules. ACM Trans. Programm. Lang. Syst. **5**(2), 190–222 (1983). https://www.microsoft.com/en-us/research/publication/specifying-concurrent-program-modules/
27. Lyons, A.: Mixed-Criticality Scheduling and Resource Sharing for High-Assurance Operating Systems. Phd thesis, University of New South Wales (Sep 2018). https://trustworthy.systems/publications/papers/Lyons%3Aphd.pdf
28. Martins, J., Pinto, S.: Shedding light on static partitioning hypervisors for arm-based mixed-criticality systems. In: 2023 IEEE 29th Real-Time and Embedded Technology and Applications Symposium (RTAS), pp. 40–53 (2023). https://doi.org/10.1109/RTAS58335.2023.00011
29. Maxim Integrated Products, Inc.: MAX15301 InTune Automatically Compensated Digital PoL Controller with Driver and PMBus Telemetry. Tech. rep. (Nov 2013)
30. Nipkow, T., Wenzel, M., Paulson, L.C.: Isabelle/HOL: a proof assistant for higher-order logic. Springer-Verlag, Berlin, Heidelberg (2002)
31. Paturel, M., Subasinghe, I., Heiser, G.: First steps in verifying the seL4 Core Platform. In: Asia-Pacific Workshop on Systems (APSys). ACM, Seoul, KR (Aug 2023). https://doi.org/10.1145/3609510.3609821
32. Pnueli, A.: The temporal logic of programs. In: 18th Annual Symposium on Foundations of Computer Science 1977 pp. 46–57 (1977). https://api.semanticscholar.org/CorpusID:117103037
33. SBS Implementers Forum: System Management Bus (SMBus) Specification (Version 2.0). Tech. rep. (Aug 2000)
34. Schirmer, N.: A sequential imperative programming language syntax, semantics, hoare logics and verification environment. Arch. Formal Proofs **2008** (2008). https://www.isa-afp.org/entries/Simpl.shtml
35. Semiconductors, N.: I2c-bus specification and user manual. Tech. rep. (Oct 2021). https://www.nxp.com/docs/en/user-guide/UM10204.pdf
36. Steven Vaughan-Nichols: MINIX: Intel's hidden in-chip operating system. https://www.zdnet.com/article/minix-intels-hidden-in-chip-operating-system/
37. Sudvarg, M., Gill, C.: A concurrency framework for priority-aware intercomponent requests in camkes on sel4. In: 2022 IEEE 28th International Conference on Embedded and Real-Time Computing Systems and Applications (RTCSA), pp. 1–10 (2022). https://doi.org/10.1109/RTCSA55878.2022.00007
38. System Management Interface Forum, Inc.: PMBus Power System Management Protocol Specification Part I – General Requirements, Transport And Electrical Interface (Revision 1.3.1). Tech. rep. (Mar 2015)
39. The seL4 community: Notifications and shared memory (Nov 2020). https://docs.sel4.systems/Tutorials/notifications.html. Accessed 28 May 2024
40. Xilinx: Zynq UltraScale+ MPSoC Software Developer Guide. Tech. Rep. UG1137 (v2022.2) (Nov 2022)

Testing Compositionality

Gijs van Cuyck[1(✉)], Lars van Arragon[1], and Jan Tretmans[1,2]

[1] Radboud University, Institute iCIS, Nijmegen, The Netherlands
{gijs.vancuyck,lars.vanarragon,jan.tretmans}@ru.nl
[2] TNO-ESI, Eindhoven, The Netherlands

Abstract. Compositionality supports the manipulation of large systems by working on their components. For model-based testing, this means that large systems can be tested by modelling and testing their components: passing tests for all components implies passing tests for the whole system. In previous work [14], we defined *mutual acceptance* for specification models and proved that this is a sufficient condition for compositionality in model-based testing. In this paper, we present an algorithm for verifying mutual acceptance on specifications and a sound and exhaustive model-based test procedure for checking mutual acceptance on black-box implementations, both inspired by the idea of *environmental conformance* [8,9]. The result is that correctness of large systems can be determined by testing the component implementations for conformance to their component specification and for environmental conformance to the specification of their environment.

1 Introduction

In recent years the amount of software that is part of everyday life has skyrocketed. However, development of these systems is only half the story, they have to be tested as well. Creating and maintaining test suites is becoming an increasingly larger part of the total cost of software development. The state-of-the-art solution to testing is called Model-Based Testing (MBT), where we employ the strengths of modelling to automatically derive tests. This lowers the testing effort from creating and maintaining test suites to creating and maintaining a model of the System Under Test (SUT). Creating small, simple models is easy, but this gets difficult quickly for larger more complicated systems. Larger models are also more difficult to understand, making it harder to transfer knowledge to new engineers and to check if the model itself is correct. Because modelling the entire system all at once is undesirable, theories have been developed that enable the composition of smaller models in order to deal with the complexity of larger systems. This comes with a catch however: if we test all of the individual components with their respective models, do we then know that the

This work is part of the project *TiCToC - Testing in Times of Continuous Change*, project nr 17936, part of the research program *MasCot - Mastering Complexity*, which is supported by the Dutch Research Council NWO.

composed system also works accordingly? To answer this question, we developed a theory for compositional model based testing called mutual acceptance ($\rightleftharpoons$) [14]. This approach of compositional testing allows to infer the correctness of a whole system from the correctness of all of its components, removing the need for manually creating large models of the entire system. In this paper, we give an algorithm to check whether two specifications are mutually accepting. We also give an algorithm to test whether a black-box implementation of a component is mutually accepting with respect to the specification of its environment. We prove that this algorithm is sound and exhaustive, and show that testing with this algorithm gives similar results as checking mutual acceptance directly on the specifications: correct components imply a correct system.

The algorithms presented in this paper are based on the theory of environmental conformance presented in [8] by Frantzen. They try to solve the same problem on a more restricted scope. In this paper we aim to combine the strong and weak points of mutual acceptance and environmental conformance into a single theory. We lift the environmental conformance relation (**eco**) to also deal with more complex label sets containing non-synchronised observable behaviour, and drop the requirement that one model must be input enabled. We show that this expanded definition of **eco** is equivalent to mutual acceptance. This allows the application of the results of our previous paper to systems that are **eco** conformant. The **eco** definition itself is also useful as it more intuitively translates to algorithms.

Overview. Section 2 contains the formal preliminaries. In Sect. 3 we lift the **eco** relation and show that **eco** implies the mutual acceptance relation and vice versa. Section 4 describes an algorithm for deciding **eco** between two specifications. We adapt this algorithm to enable on-the-fly testing of **eco** between an implementation and (part of) its environment specification. We conclude this section by showing that correctness is preserved over parallel composition when testing under **eco**. In Sect. 5 we formalise the in the previous section described testing and prove our algorithm correct. Sections 6 to 8 contain the related work, future work, and conclusion respectively. All proofs can be found in [3].

2   Preliminaries

This section contains the theoretical background for our work. It is a shortened version of the preliminaries and lemmas from [14], which is based on [13].

The main formalism used is that of labelled transition systems (LTS) (Definition 1). An LTS has states and transitions between states that model events. An event can be an observable input or output, or an unobservable internal transition τ. I_s, U_s, etc., indicate inputs and outputs, respectively, coming from LTS s. The shorthand L_s means $I_s \cup U_s$. The name of an LTS is sometimes used as shorthand for its starting state. $\mathcal{LTS}$ denotes the domain of labelled transition systems. For technical reasons we restrict this class to strongly converging and image-finite systems. Strong convergence means that infinite sequences of

τ-actions are not allowed to occur. Image-finiteness means that the number of non-deterministically reachable states shall be finite. In examples, input and output sets are given implicitly by prefixing inputs with ?, and outputs with !. The same label can be in the input set of one LTS and in the output set of another. $X \in \mathcal{P}(Q)$ denotes a subset of Q.

Definition 1. *A* Labelled Transition System *is a 5-tuple* $\langle Q, I, U, T, q_0 \rangle$ *where: Q is a non-empty, countable set of states; I is a countable set of input labels; U is a countable set of output labels, which is disjoint from I; $T \subseteq Q \times (I \cup U \cup \{\tau\}) \times Q$ is a set of triples, the transition relation; $q_0 \in Q$ is the initial state.*

Reasoning about labelled transition systems uses the concept of traces. A trace is a sequence of observable labels that can occur when walking through an LTS. Common notation used when describing traces is repeated in Definition 2.

Definition 2. *Let* $s \in \mathcal{LTS}$; $q, q' \in Q_s$; $\ell \in L_s$; $\sigma \in L_s^*$; $\ell_\tau \in L_s \cup \{\tau\}$; $\sigma_\tau \in (L_s \cup \{\tau\})^*$, *where* ϵ *denotes the empty sequence of labels.*

$$
\begin{aligned}
q \xrightarrow{\epsilon} q' & \overset{def}{=} q = q' \\
q \xrightarrow{\ell_\tau} q' & \overset{def}{=} (q, \ell_\tau, q') \in T_s \\
q \xrightarrow{\ell_\tau \cdot \sigma_\tau} q' & \overset{def}{=} \exists q'' \in Q_s : q \xrightarrow{\ell_\tau} q'' \wedge q'' \xrightarrow{\sigma_\tau} q' \\
q \xrightarrow{\sigma} & \overset{def}{=} \exists q'' \in Q_s : q \xrightarrow{\sigma} q'' \\
q \not\xrightarrow{\sigma} & \overset{def}{=} \nexists q'' \in Q_s : q \xrightarrow{\sigma} q'' \\
q \overset{\epsilon}{\Rightarrow} q' & \overset{def}{=} \exists \varphi \in \{\tau\}^* : q \xrightarrow{\varphi} q' \\
q \overset{\sigma \cdot \ell}{\Longrightarrow} q' & \overset{def}{=} \exists q'', q''' \in Q_s : q \overset{\sigma}{\Rightarrow} q'' \wedge q'' \xrightarrow{\ell} q''' \wedge q''' \overset{\epsilon}{\Rightarrow} q'
\end{aligned}
$$

While specifications are often given as an LTS, IOTS are used to represent implementations. In MBT, we assume that we can always give any input to an implementation. $\mathcal{IOTS}$ denotes the domain of all input-enabled transition systems.

Definition 3. $i \in \mathcal{LTS}$ *is called an* Input-Enabled Transition System *(IOTS) if in every state for every input, its transition relation either defines that input, or reaches with just internal transitions another state that does so:*

$$
\forall q \in Q_i, \; \ell \in I_i : \; q \overset{\ell}{\Rightarrow}
$$

Multiple labelled transition systems can be composed using parallel composition (Definition 5) to form larger models. The result of parallel composition represents a system where all the components are being executed at the same time independently of each other. Synchronous communication occurs on shared labels. For brevity, we implicitly assume $s, e \in \mathcal{LTS}$, and s and e are *composable*.

Definition 4. $s, e \in \mathcal{LTS}$ *are* composable *iff their respective output sets* U_s *and* U_e *are disjoint:* $U_s \cap U_e = \emptyset$

Definition 5. *Parallel composition* $\parallel$ *on two composable labelled transition systems s and e is defined as* $\quad s \parallel e \overset{def}{=} \langle Q, I, U, T, q_0 \rangle$, *where:*
$Q = \{q_s \parallel q_e \mid q_s \in Q_s, q_e \in Q_e\}$; $I = (I_s \setminus U_e) \cup (I_e \setminus U_s)$; $U = U_s \cup U_e$;
$q_0 = q_{0_s} \parallel q_{0_e}$; T *is the minimal set satisfying the following inference rules (where $q_s, q_s' \in Q_s, q_e, q_e' \in Q_e$):*

$$
\begin{array}{llll}
q_s \xrightarrow{\ell} q_s' & \ell \in (L_s \cup \{\tau\}) \setminus L_e & \vdash & q_s \parallel q_e \xrightarrow{\ell} q_s' \parallel q_e \\
q_e \xrightarrow{\ell} q_e' & \ell \in (L_e \cup \{\tau\}) \setminus L_s & \vdash & q_s \parallel q_e \xrightarrow{\ell} q_s \parallel q_e' \\
q_s \xrightarrow{\ell} q_s', \; q_e \xrightarrow{\ell} q_e' & \ell \in L_s \cap L_e & \vdash & q_s \parallel q_e \xrightarrow{\ell} q_s' \parallel q_e'
\end{array}
$$

A conformance relation describes when an implementation is considered correct according to its specification. We use the **uioco** conformance relation in this paper [15]. This means implementations are correct if their outputs in states reachable by *Utraces* are a subset of what the specification allows. *Utraces* are traces through the specification, while only allowing inputs if they are enabled in all possible current states. This is formalised in Definitions 6 to 8.

Definition 6. $\delta \notin L_s$ *is a special output denoting the absence of outputs, called quiescence. It is defined as follows (with $q_s, q_s' \in Q_s$):*

$$
q_s \xrightarrow{\delta} q_s' \overset{def}{=} q_s = q_s' \;\wedge\; \forall x \in U_s \cup \{\tau\}: \; q_s \not\xrightarrow{x}
$$

L^δ *is used as shorthand for $L \cup \{\delta\}$.*

Definition 7. *Let $q_s \in Q_s$; $Q \subseteq Q_s$ and $\sigma \in L_s^{\delta *}$.*

$$
\begin{array}{ll}
q_s \textbf{ after } \sigma & \overset{def}{=} \{ q_s' \in Q_s \mid q_s \xRightarrow{\sigma} q_s' \} \\
\textbf{out}(q_s) & \overset{def}{=} \{ x \in U_s^\delta \mid q_s \xrightarrow{x} \} \\
\textbf{out}(Q) & \overset{def}{=} \bigcup \{ \textbf{out}(q) \mid q \in Q \} \\
\textbf{in}(q_s) & \overset{def}{=} \{ \ell \in I_s \mid q_s \xRightarrow{\ell} \} \\
\textbf{in}(Q) & \overset{def}{=} \bigcap \{ \textbf{in}(q) \mid q \in Q \}
\end{array}
$$

Definition 8. *Let $i \in \mathcal{IOTS}$, with $I_i = I_s$ and $U_i = U_s$:*

$$
Utraces(s) \overset{def}{=} \{ \sigma \in L^{\delta *} \mid s \xRightarrow{\sigma} \wedge \, (\forall q_s \in Q_s, \sigma_1 \cdot \ell \cdot \sigma_2 = \sigma, \ell \in I :
$$
$$
s \xRightarrow{\sigma_1} q_s \implies q_s \xRightarrow{\ell}) \}
$$
$$
i \textbf{ uioco } s \overset{def}{=} \forall \sigma \in Utraces(s): \; \textbf{out}(i \textbf{ after } \sigma) \subseteq \textbf{out}(s \textbf{ after } \sigma)
$$

To relate traces from labelled transition systems with different label sets with each other, we often have to remove or replace certain labels from traces. This is done using projection and substitution defined in Definitions 9 and 10.

Definition 9. *Let $\sigma \in L^{\delta *}$; $\ell \in L^\delta$; and $\mathcal{L}$ be any set of labels. Projecting a trace to a restricted set of labels is defined as:*

$$
\begin{array}{ll}
\epsilon \restriction \mathcal{L} & \overset{def}{=} \epsilon \\
(\sigma \cdot \ell) \restriction \mathcal{L} & \overset{def}{=} (\sigma \restriction \mathcal{L}) \cdot \ell \quad \text{if } \ell \in \mathcal{L} \\
& \qquad \sigma \restriction \mathcal{L} \qquad\quad otherwise
\end{array}
$$

Definition 10. *Let* $\sigma \in L^{\delta*}$; $\ell, \ell'' \in L^{\delta}$ *with* $\ell \neq \ell''$; $\mathcal{L}$ *be any set of labels,* $\ell' \in \mathcal{L}$. *Then* substitutution *defines another trace with some labels exchanged with other labels:*

$$\epsilon[\ell \mapsto \ell'] \quad \overset{def}{=} \quad \epsilon$$
$$(\sigma \cdot \ell)[\ell \mapsto \ell'] \quad \overset{def}{=} \quad \sigma[\ell \mapsto \ell'] \cdot \ell'$$
$$(\sigma \cdot \ell)[\ell'' \mapsto \ell'] \quad \overset{def}{=} \quad \sigma[\ell'' \mapsto \ell'] \cdot \ell$$

Parallel composition does not preserve correctness under **uioco** in general: i_s **uioco** $s \wedge i_e$ **uioco** $e \not\Longrightarrow i_s \parallel i_e$ **uioco** $s \parallel e$. However, it does preserve this correctness for specifications that are mutually accepting (Definition 11), which is the main result of our previous work (Theorem 1) [14]. This means that when an output is produced for another system, this other system must accept this output as input, whenever this output can be produced. This is the basis we use to further develop compositional model based testing.

Definition 11. *s* *accepts(↪) / mutually accepts(⇋) e, respectively, iff:*

$$s \hookleftarrow e \quad \overset{def}{=} \quad \forall \sigma \in Utraces(s \parallel e), \ q_s \in Q_s, \ q_e \in Q_e :$$
$$s \parallel e \overset{\sigma}{\Rightarrow} q_s \parallel q_e \implies \boldsymbol{out}(q_e) \cap I_s \subseteq \boldsymbol{in}(q_s) \cap U_e$$
$$s \leftrightharpoons e \quad \overset{def}{=} \quad s \hookleftarrow e \ \wedge \ e \hookleftarrow s$$

Theorem 1. *Let* $i_s, i_e \in \mathcal{IOTS}$, *then*

$$s \leftrightharpoons e \ \wedge \ i_s \ \boldsymbol{uioco} \ s \ \wedge \ i_e \ \boldsymbol{uioco} \ e \implies i_s \parallel i_e \ \boldsymbol{uioco} \ s \parallel e$$

3   Environmental Conformance

Mutually accepting systems are useful for model based testing, because they preserve correctness: **uioco** correct components imply the **uioco** correctness of their parallel composition. The definition used here and in [14] however does not intuitively lead to an algorithm. Earlier work on environmental conformance (**eco**) does have some algorithms [8], but the more limited scope does not make these algorithms directly applicable. In this section we formalise a more general version of the **eco** relation defined in [8]. We then go on to prove that this new definition coincides with ⇋.

Definitions 12 and 13 adapt the idea of an **eco**-*bisimulation* from [8] in several ways. We start by adding internal transitions, which are not synchronised between the two components. This means that any component can do any number of them at any time without the other component being aware. Next, we add synchronised inputs to account for our expanded definition of *composable* (Definition 4). Additionally, we have to take quiescence into account. This is required to make the link between **eco** conformance of components and **uioco** conformance of the composed system. Finally, we drop the requirement that one of the two models is input enabled, which allows for the use of model checking to prove **eco** conformance directly on specifications.

Definition 12. $\mathcal{R} \subseteq \mathcal{P}(Q_s) \times \mathcal{P}(Q_e)$ *is an* **eco-bisimulation** *for s and e iff:*

1. $(s \; \textbf{after} \; \epsilon, e \; \textbf{after} \; \epsilon) \in \mathcal{R}$
2. $\forall (X_s, X_e) \in \mathcal{R}$:

   (a) $\forall \ell \in \textbf{out}(X_e) \cap I_s : \ell \in \textbf{in}(X_s) \wedge (X_s \; \textbf{after} \; \ell, X_e \; \textbf{after} \; \ell) \in \mathcal{R}$

   (b) $\forall \ell \in \textbf{out}(X_s) \cap I_e : \ell \in \textbf{in}(X_e) \wedge (X_s \; \textbf{after} \; \ell, X_e \; \textbf{after} \; \ell) \in \mathcal{R}$

   (c) $\forall \ell \in (\textbf{out}(X_e) \cup \textbf{in}(X_e)) \setminus L_s^\delta : (X_s, X_e \; \textbf{after} \; \ell) \in \mathcal{R}$

   (d) $\forall \ell \in (\textbf{out}(X_s) \cup \textbf{in}(X_s)) \setminus L_e^\delta : (X_s \; \textbf{after} \; \ell, X_e) \in \mathcal{R}$

   (e) $\forall \ell \in (\textbf{in}(X_s) \cap \textbf{in}(X_e)) : (X_s \; \textbf{after} \; \ell, X_e \; \textbf{after} \; \ell) \in \mathcal{R}$

   (f) $\delta \in \textbf{out}(X_s) \wedge \delta \in \textbf{out}(X_e) \implies (X_s \; \textbf{after} \; \delta, X_e \; \textbf{after} \; \delta) \in \mathcal{R}$

Definition 13. *s is* **eco**-*conform to e, denoted s* **eco** *e, iff there exists an* **eco**-*bisimulation for s and e.*

Lemma 1.

$$s \; \textbf{eco} \; e \iff e \; \textbf{eco} \; s$$

An **eco**-*bisimulation* relates the sets of states the components of a composed system can be in. It then adds the requirement that synchronised outputs generated by one component have to be defined as inputs in all corresponding reachable states of the other component, similar to mutual acceptance.

Points 2a and 2b cover the synchronised outputs. If they happen, both components change states. If an output is communicated, the receiving component must define how to handle it as an input. Points 2c and 2d cover the unsynchronised labels which only happen in one component. Point 2e covers synchronised inputs that are not outputs of the other component. These come from an unspecified third component that is part of the new environment after composition. These are only relevant if they are enabled in both s and e, because otherwise they are not part of the *Utraces* of the composed system. The last point, 2f, covers quiescence. This is relevant here because if the system displays quiescence, this is observable and it reduces the uncertainty over which state the system is in. Quiescence is only observable in a composed system if all components are quiescent, so cases were only one component is quiescent can be ignored here.

Definition 14. $\mathcal{R}_U(s, e) \subseteq \mathcal{P}(Q_s) \times \mathcal{P}(Q_e)$, *is the minimal set satisfying the following inference rules:*

$$\vdash (s \; \textbf{after} \; \epsilon, e \; \textbf{after} \; \epsilon) \in \mathcal{R}_U \qquad (1)$$

$$(X_s, X_e) \in \mathcal{R}_U, \qquad \ell \in \textbf{out}(X_e) \cap I_s \vdash (X_s \; \textbf{after} \; \ell, X_e \; \textbf{after} \; \ell) \in \mathcal{R}_U \qquad (2)$$

$$(X_s, X_e) \in \mathcal{R}_U, \qquad \ell \in \textbf{out}(X_s) \cap I_e \vdash (X_s \; \textbf{after} \; \ell, X_e \; \textbf{after} \; \ell) \in \mathcal{R}_U \qquad (3)$$

$$(X_s, X_e) \in \mathcal{R}_U, \qquad \ell \in (\textbf{out}(X_e) \cup \textbf{in}(X_e)) \setminus L_s^\delta \vdash (X_s, X_e \; \textbf{after} \; \ell) \in \mathcal{R}_U \qquad (4)$$

$$(X_s, X_e) \in \mathcal{R}_U, \qquad \ell \in (\textbf{out}(X_s) \cup \textbf{in}(X_s)) \setminus L_e^\delta \vdash (X_s \; \textbf{after} \; \ell, X_e) \in \mathcal{R}_U \qquad (5)$$

$$(X_s, X_e) \in \mathcal{R}_U, \qquad \ell \in (\textbf{in}(X_s) \cap \textbf{in}(X_e)) \vdash (X_s \; \textbf{after} \; \ell, X_e \; \textbf{after} \; \ell) \in \mathcal{R}_U \qquad (6)$$

$$(X_s, X_e) \in \mathcal{R}_U, \qquad \delta \in \textbf{out}(X_s) \wedge \delta \in \textbf{out}(X_e) \vdash (X_s \; \textbf{after} \; \delta, X_e \; \textbf{after} \; \delta) \in \mathcal{R}_U \qquad (7)$$

where $X_s \subseteq Q_s$, $X_e \subseteq Q_e$, $\ell \in L_s \cup L_e \cup \{\delta\}$

Definition 13 defines s **eco** e as the existence of an **eco**-*bisimulation*. This exactly defines the idea of environmental conformance, but because it does not exclude unnecessary elements it is difficult to work with. We therefore also introduce a concrete example of such a relation in Definition 14, which represents the 'simplest' **eco**-*bisimulation* (Lemmas 2 and 3). This means that if an **eco**-*bisimulation* exists for s and e, then $\mathcal{R}_U(s,e)$ relates the fewest possible elements while still being an **eco**-*bisimulation*. We write $\mathcal{R}_U$ for $\mathcal{R}_U(s,e)$ when s and e are clear from context.

Lemma 2. *Let $\mathcal{R}$ be an **eco**-bisimulation for s and e, then:*

$$\mathcal{R}_U \subseteq \mathcal{R}$$

Lemma 3.

$$s \ \boldsymbol{eco} \ e \iff \mathcal{R}_U \ \text{is an } \boldsymbol{eco}\text{-bisimulation for } s \text{ and } e$$

3.1   Eco Implies Mutual Acceptance

In order to connect the algorithms available for **eco** to the theoretical results for mutual acceptance, we show that s **eco** e implies $s \leftrightharpoons e$. For states within an **eco**-*bisimulation* $\mathcal{R}$, quiescence is preserved over parallel composition, which is not the case in general. Since *Utraces* without δ are preserved under parallel composition, this generalises to all *Utraces* in Lemma 4. However, just showing that the traces are preserved under composition is not enough on its own. We must also show the states within an **eco**-*bisimulation* are exactly those reachable by *Utraces* (Lemma 5).

Lemma 4. *Let $\sigma \in Utraces(s \,||\, e)$, $q_s \in Q_s$, $q_e \in Q_e$:*

$$s \ \boldsymbol{eco} \ e \implies (s \,||\, e \overset{\sigma}{\Rightarrow} q_s \,||\, q_e \iff s \xrightarrow{\sigma \restriction L_s^\delta} q_s \land e \xrightarrow{\sigma \restriction L_e^\delta} q_e)$$

Lemma 5. *Let s **eco** e, $q_s \in Q_s$, $q_e \in Q_e$.*

$$\exists (X_s, X_e) \in \mathcal{R}_U : q_s \in X_s \land q_e \in X_e \iff \exists \sigma \in Utraces(s \,||\, e) : s \,||\, e \overset{\sigma}{\Rightarrow} q_s \,||\, q_e$$

Using these two results, we can now prove that the existence of an **eco**-*bisimulation* for s and e implies certain properties over the states reachable by *Utraces* in the parallel composition of s and e. Most notably, we can conclude from Definition 12.2a and Definition 12.2b that synchronised outputs of one component are always accepted by the other component, which coincides with the definition of $\leftrightharpoons$. This means that establishing the existence of an **eco**-*bisimulation* is enough to apply the theoretical results previously obtained for $\leftrightharpoons$ in [14], the main result being the possibility of compositional testing as expressed in Theorem 3.

Theorem 2.
$$s \; \textbf{\textit{eco}} \; e \implies s \leftrightarrows e$$

Theorem 3. *Let* $i_s, i_e \in \mathcal{IOTS}$.
$$s \; \textbf{\textit{eco}} \; e \wedge i_s \; \textbf{\textit{uioco}} \; s \wedge i_e \; \textbf{\textit{uioco}} \; e \implies i_s \, \| \, i_e \; \textbf{\textit{uioco}} \; s \, \| \, e$$

3.2   Mutual Acceptance Implies Eco

The previous section shows that **eco** is at least as strong as $\leftrightarrows$. In this section we show that the other direction also holds, and that **eco** and $\leftrightarrows$ are just two different ways of writing down the same relation. This means **eco** is not just sufficient, but also necessary with respect to mutual acceptance, and that they can be used interchangeably wherever this is convenient.

To prove this, we show that under the assumption of $s \leftrightarrows e$, Definitions 12 and 14 coincide (Lemma 7). The proof follows the same structure and lemmas as the previous section, but steps relying on Definition 12.2a and Definition 12.2b are replaced with steps relying on $s \leftrightarrows e$ instead. Combined with Theorem 2, this then shows that s **eco** e and $s \leftrightarrows e$ are equivalent in Theorem 4.

Lemma 6. *Let* $s \leftrightarrows e$, $q_e \in Q_e, q_s \in Q_s$.

$$\exists (X_s, X_e) \in \mathcal{R}_U : q_s \in X_s \wedge q_e \in X_e \implies \exists \sigma \in Utraces(s \, \| \, e) : s \, \| \, e \xRightarrow{\sigma} q_s \, \| \, q_e$$

Lemma 7.
$$s \leftrightarrows e \implies \mathcal{R}_U \; \text{is an } \textbf{eco}\text{-bisimulation}$$

Theorem 4.
$$s \; \textbf{\textit{eco}} \; e \iff s \leftrightarrows e$$

4   Testing Accepting Systems

The previous sections described why the environmental conformance and mutually accepting relations are the same. In this section we will make use of this by giving an algorithm for checking environmental conformance, and by extension also mutual acceptance. Because environmental conformance comes down to the existence of an **eco**-*bisimulation*, the algorithm works by constructing such a relation. It is an adapted version of the algorithm introduced in [8].

DecideEco and Algorithm 1 show a recursive function that will check whether two specifications are environmentally conformant. This works by constructing the **uioco**-*bisimulation* from Definition 14, and then checking if it is an **eco**-*bisimulation*. We use the fact that if an **eco**-*bisimulation* exists, then the **uioco**-*bisimulation* is such a relation (Lemma 3).

Function DecideEco($(X_s, X_e) : \mathcal{P}(Q_s) \times \mathcal{P}(Q_e)$)

Data: $\mathcal{R} : Set\langle \mathcal{P}(Q_s), \mathcal{P}(Q_e)\rangle$

```
 1: if out(X_s) ∩ I_e ⊆ in(X_e) ∧ out(X_e) ∩ I_s ⊆ in(X_s) then
 2: |   R ← R ∪ {(X_s, X_e)};
 3: |   A ← (out(X_s) ∩ I_e) ∪ (out(X_e) ∩ I_s) ∪ (in(X_s) ∩ in(X_e));
 4: |   if δ ∈ out(X_e) ∧ δ ∈ out(X_s) then
 5: |   |   A ← A ∪ {δ};
 6: |   foreach ℓ ∈ A do
 7: |   |   next ← (X_s after ℓ, X_e after ℓ);
 8: |   |   if next ∉ R then
 9: |   |   |   DecideEco(next)
10: |   foreach ℓ ∈ (out(X_s) ∪ in(X_s)) \ L_e^δ do
11: |   |   next ← (X_s after ℓ, X_e);
12: |   |   if next ∉ R then
13: |   |   |   DecideEco(next)
14: |   foreach ℓ ∈ (out(X_e) ∪ in(X_e)) \ L_s^δ do
15: |   |   next ← (X_s, X_e after ℓ);
16: |   |   if next ∉ R then
17: |   |   |   DecideEco(next)
18: else
19: |   throw exception s ̸≡eco e
```

Algorithm 1: Deciding eco

Input: $s, e : \mathcal{LTS}$ with finite number of states and labels

```
1: R ← ∅;
2: DecideEco((s_0 after ε, e_0 after ε))
```

Since constructing $\mathcal{R}$ means exploring the part of the state space that is reachable by *Utraces*, the algorithm will only terminate if this state space is finite. The number of enabled labels in a given state q: $\mathbf{out}(q) \cup \mathbf{in}(q)$, must also be finite.

The correctness of the algorithm follows directly from Definitions 12 and 13. The algorithm constructs an **eco**-*bisimulation* by adding the minimum amount of elements to the relation as required. If it fails in doing so, which can only happen by failing the conditions of either Definition 12.2a or Definition 12.2b, then the two specifications are not environmentally conformant. For simplicity Algorithm 1 does not give a counterexample and stops at the first found problem. This could however be added by also keeping track of the sequence of labels that was used to reach a certain set of states, at the cost of increased memory consumption.

4.1   On-the-Fly Testing

Algorithm 1 can be used to check **eco** conformance between two specifications. Sometimes this might not be feasible however, for instance when flattening symbolic specifications containing data. This will often result in both an infinite state space and an infinite label set, which means Algorithm 1 will not terminate. In those cases, we can still use testing to get some idea on the environmental conformance between an implementation and a specification of its environment.

This can be done with Algorithm 2, which is an adapted version of the algorithm from [8]. Intuitively the algorithm is very similar to the algorithm used for testing **uioco** conformance and consists of five possible options which repeat non-deterministically until the algorithm terminates: we stop testing and the test passes; we give a valid input to the SUT, based on which outputs the environment is currently allowed to send; we can observe an output from the SUT, and check if the environment can handle this output; we can simulate internal behaviour of the environment; or we can generate inputs for the SUT not coming from the environment. This last option can be optimised further if a specification of s is available, by restricting to specified inputs instead of random ones. We claim the algorithm is sound and complete. This means that it is possible to generate a test that will fail, if and only if the SUT and its environment are not **eco** conformant. We will further formalise and prove this claim in Sect. 5.

Algorithm 2: On-the-Fly Testing for **eco**

Input: $e \in \mathcal{LTS}$, connection to SUT, L_s: labels of SUT

1: $X_e \leftarrow e_0$ **after** ϵ;
    nondeterministically execute a finite number of the following cases,
    until the test either Passes or Fails:
    (A) *Stop testing:*
2:      the test Passes;
    (B) *Emit a response to the SUT:*
3:      choose $\ell \in (\mathbf{out}(X_e) \cup \mathbf{in}(X_e)) \cap I_s$;
4:      send ℓ to the SUT;
5:      $X_e \leftarrow X_e$ **after** ℓ;
    (C) *Observe a request ℓ from the SUT:*
6:      **if** $\ell \in I_e$ **then**
7:          **if** $\ell \notin \mathbf{in}(X_e)$ **then** the test Fails;
            else
8:          $\quad |\quad X_e \leftarrow X_e$ **after** ℓ;
        else if $\ell = \delta \wedge \delta \in \mathbf{out}(X_e)$ **then**
9:          $| \quad X_e \leftarrow X_e$ **after** ℓ;
    (D) *Simulate internal behaviour in the environment*
10:     choose $\ell \in (\mathbf{out}(X_e) \cup \mathbf{in}(X_e)) \setminus L_s$;
11:     $X_e \leftarrow X_e$ **after** ℓ;
    (E) *Generate internal behaviour of the SUT*
12:     choose $\ell \in I_s \setminus L_e$;
13:     send ℓ to the SUT;

4.2 Component Based Testing

Using the algorithm introduced in the previous section, we can test for environmental conformance between an implementation and a specification of its environment ($i_s \leftrightharpoons e$). This however still leaves the question if it is also useful. Does black-box testing using an implementation give us enough information to draw the same conclusions as if the specifications were mutually accepting? Environmental conformance for an implementation i_s and its environment specification e does not imply conformance between the specification s and its environment: $i_s \leftrightharpoons e \wedge i_s$ **uioco** $s \not\Rightarrow s \leftrightharpoons e$. This does not mean that testing for environmental conformance is not useful. It still gives enough information to guarantee correctness of compositional testing, which is formulated in Theorem 5, and visualised in Fig. 1. Testing the components for uioco correctness is represented by the solid arrows and is always required. Then, in addition to that we either need to check environmental conformance on the specifications (Theorem 3) [14] represented by the horizontal arrow, or test this on the implementations instead using the diagonal arrows (Theorem 5). Both options then result in the conclusion that the composed implementation is also uioco conforming to the composed specification.

Theorem 5. *Let $i_s, i_e \in \mathcal{IOTS}$*

$$i_s \leftrightharpoons e \wedge i_e \leftrightharpoons s \wedge i_s \ \textbf{\textit{uioco}} \ s \wedge i_e \ \textbf{\textit{uioco}} \ e \implies i_s \parallel i_e \ \textbf{\textit{uioco}} \ s \parallel e$$

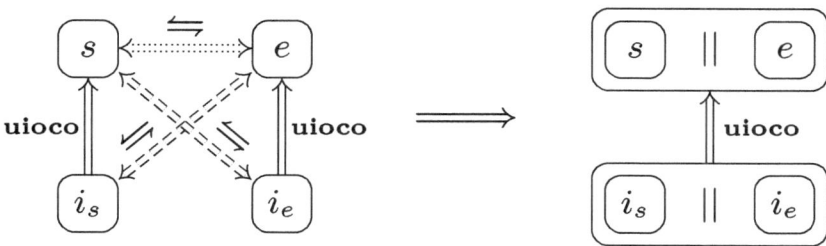

Fig. 1. Testing relations overview

5   Testing Formalisation

In previous sections we have introduced an algorithm for testing environmental correctness between an implementation and a specification of its environment (Algorithm 2), and shown how using this algorithm allows compositional testing (Theorem 5). The correctness of this algorithm has however been left implicit, partially because of the ambiguous nature of pseudocode describing communication. In this section, we formalise what it means for a testing algorithm to be correct without relying on ambiguous pseudocode. We define a functional variant of Algorithm 2, and then prove this new algorithm correct.

5.1   Background

We start with providing the definitions used to formalise test cases, test execution and correctness. The content of this section is an adaptation of the test case formalisation used in [9,13].

A test case t (Definition 17) is a special LTS with finite behaviour that describes a test to be executed on an implementation (the SUT). A test case describes interactions with the SUT, therefore the inputs of a test case are the outputs of the SUT, and vice versa. Actions given to the SUT have to be deterministic. In addition to this, a test case has one special input θ, which represents observing quiescence from the SUT. A test case also has two special states, **pass** and **fail**, which represent the test passing or failing respectively.

In order to recursively define larger tests from smaller tests, we use behaviour expressions [13] (Definition 15), which are written in a form of process algebra originally based on LOTOS [2]. The semantics of behaviour expressions are given in terms of labelled transition systems in Definition 16. In our case behaviour expressions can consist of either a single transition followed by more behaviour or choice over a possibly empty set of behaviour expressions.

Definition 15. *A **behaviour expression** uses the following syntax:*

$$B ::= \quad \ell\,;B \quad | \quad \Sigma \mathcal{B}$$

Where ℓ is a label, $\mathcal{B}$ is a set of behaviour expressions. We also use $B_1 \,\#\# \, B_2$ as shorthand for $\Sigma\{B_1, B_2\}$.

Definition 16. *A behaviour expression represents (a state in) a labelled transition system, with the transition relation given by the following inference rules:*

$$\vdash \ \ell\,;B \xrightarrow{\ell} B$$
$$B \in \mathcal{B}, \quad B \xrightarrow{\ell} B' \quad \vdash \ \Sigma\mathcal{B} \xrightarrow{\ell} B'$$

Definition 17. *A **test case** t is an LTS with the following properties:*

1. *t is a finite tree: Every non-terminal state except the starting state has exactly one incoming transition.*
2. *t has two terminal states: **pass** and **fail**, which have self loops for every input label: $\textbf{fail} ::= \Sigma\{\ell\,;\textbf{fail} \mid \ell \in I_t\}$ and $\textbf{pass} ::= \Sigma\{\ell\,;\textbf{pass} \mid \ell \in I_t\}$*
3. *t can react to quiescence, using the special input $\theta \in I_t$.*
4. *Every state of t has at most one outgoing transition from U_t.*
5. *t is receptive: every state has exactly one outgoing input transition for every $\ell \in I_t \setminus \{\theta\}$.*
6. *t is fully observable: there are no internal transitions τ.*

The domain of all test cases is denoted as $\mathcal{TTS}$, and a collection of test cases is called a test suite: $TS \subseteq \mathcal{TTS}$.

During test execution, the test case takes the role of the environment in a modified form of parallel composition (Definition 18). The inputs and outputs of the test case are mostly the inverse of the SUT, but not exactly. In Fig. 2, arrows 1, 3, 5, 6 and 7 are decisions made by the testcase and therefore outputs.

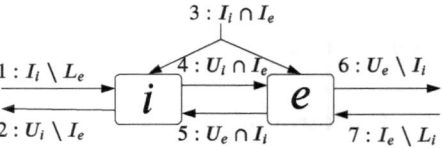

Fig. 2. Labels of composed system

Arrow 4 and δ are decided by the SUT and therefore inputs. Arrow 2 is not needed for determining **eco** conformance and is not part of testing here.

Definition 18. *Let $t \in \mathcal{TTS}$, $i \in \mathcal{IOTS}$. Then **test execution** $t \,]\!|\, i$ is defined as a variant on $||$. $Q_{t]|i} = Q_t \times Q_i$, $U_{t]|i} = U_i \cup L_t$, $I_{t]|i} = \emptyset$. The transition function $T_{t]|i}$ is defined as the minimal set satisfying the following inference rules (where $q_t, q_t' \in Q_t$, $q_i, q_i' \in Q_i$):*

$$\frac{q_i \xrightarrow{\tau} q_i'}{} \qquad\qquad \vdash \quad q_t \,]\!|\, q_i \xrightarrow{\tau} q_t \,]\!|\, q_i'$$

$$\frac{q_i \xrightarrow{\ell} q_i', \quad \ell \in L_i \setminus L_t}{} \qquad \vdash \quad q_t \,]\!|\, q_i \xrightarrow{\ell} q_t \,]\!|\, q_i'$$

$$\frac{q_t \xrightarrow{\ell} q_t', \qquad\qquad\qquad \ell \in L_t \setminus (L_i \cup \{\theta\})}{} \quad \vdash \quad q_t \,]\!|\, q_i \xrightarrow{\ell} q_t' \,]\!|\, q_i$$

$$\frac{q_t \xrightarrow{\ell} q_t', \quad q_i \xrightarrow{\ell} q_i', \quad \ell \in L_i \cap L_t}{} \quad \vdash \quad q_t \,]\!|\, q_i \xrightarrow{\ell} q_t' \,]\!|\, q_i'$$

$$\frac{q_t \xrightarrow{\theta} q_t', \quad q_i \xrightarrow{\delta} q_i'}{} \qquad\qquad \vdash \quad q_t \,]\!|\, q_i \xrightarrow{\theta} q_t' \,]\!|\, q_i'$$

A test run of a test case t for an implementation i is a trace through its test execution $t \,]\!|\, i$ which ends in a terminal state. An implementation i *fails* a test case t if there exists a test run that reaches **fail**. Conversely, an implementation *passes* a test case if all possible test runs reach **pass**.

Definition 19. *Let $t \in \mathcal{TTS}$, $i \in \mathcal{IOTS}$.*

1. $\sigma \in L^*_{t]|i}$ *is a **test run** for t on i if it reaches a terminal state:*

$$\exists q_i \in Q_i : \quad t \,]\!|\, i \xRightarrow{\sigma} \textbf{pass} \,]\!|\, q_i \quad \vee \quad t \,]\!|\, i \xRightarrow{\sigma} \textbf{fail} \,]\!|\, q_i$$

   ***Runs**($t \,]\!|\, i$) represents the set of all test runs for t on i*
2. i *passes t if none of the test runs of t for i reaches the **fail** state:*

$$i \text{ passes } t \stackrel{def}{=} \forall \sigma \in \textbf{Runs}(t \,]\!|\, i), \forall q_i \in Q_i : t \,]\!|\, i \xRightarrow{\sigma}\!\!\!\!\!/\;\; \textbf{fail} \,]\!|\, q_i$$

3. i *passes $TS \subseteq \mathcal{TTS}$ if all the tests in TS pass:*

$$i \text{ passes } TS \stackrel{def}{=} \forall t \in TS : i \text{ passes } t$$

4. *if i does not pass a test case or test suite, it fails*

Using these properties we can define what it means for a test suite to be correct in Definition 20. A test suite is *sound* when it only fails for implementations that are not **eco** correct with respect to the given environment specification. A test suite is *exhaustive* if for every implementation that is not **eco** correct, there is a test in the test suite that will fail for this implementation. A test suite is *complete* if it has both of these properties at the same time. In practice, an exhaustive test suite consists of an infinite amount of tests, and any practical application of the test suite will have to do some test selection. It can nevertheless be useful to start with an exhaustive test suite, such that for every fault it is at least possible to select a test which will find that fault.

Definition 20. *Let* $TS \subseteq TTS$:

$$
\begin{array}{llll}
TS \text{ is } \textbf{complete} & \stackrel{def}{=} & \forall i \in \mathcal{IOTS} : i \text{ } eco \text{ } e & \Longleftrightarrow & i \text{ } passes \text{ } TS \\
TS \text{ is } \textbf{sound} & \stackrel{def}{=} & \forall i \in \mathcal{IOTS} : i \text{ } eco \text{ } e & \Longrightarrow & i \text{ } passes \text{ } TS \\
TS \text{ is } \textbf{exhaustive} & \stackrel{def}{=} & \forall i \in \mathcal{IOTS} : i \text{ } eco \text{ } e & \Longleftarrow & i \text{ } passes \text{ } TS
\end{array}
$$

5.2 Algorithm Correctness

This section will give arguments for the correctness of Algorithm 2. It is difficult to prove the correctness of an imperative and interactive program. We make this simpler by defining a non-interactive variant of the algorithm (Algorithm 3) which generates a test case instead, and proving this variant to be correct.

The main difference with Algorithm 2 is that instead of interacting with the SUT, it non-deterministically generates a test case. Running this test case with the SUT then corresponds to a possible run of Algorithm 2. Algorithm 3 covers the same nondeterministic cases as Algorithm 2. A difference is that Algorithm 2 implicitly assumes it can switch to handling an input from the *SUT* if it receives one. Algorithm 3 makes this explicit by adding transitions for things that could happen outside of the control of the test case to every state, using the `GenInputResponse` function.

By running Algorithm 3 multiple times, we can generate a test suite. We denote the test suite generated by Algorithm 3 for a label set L_i and environment specification e as $TTS(L_i, e)$. We prove that the test suite generated by Algorithm 3 is complete. This means that if an implementation is not **eco** conformant to a specification, then Algorithm 3 can generate a test case that can fail when executed on that implementation. Conversely, none of the test cases generated can fail on an implementation that is **eco** conformant.

The proof mostly involves rewriting traces between the various transition systems involved, as expressed in Lemmas 8 to 10. This uses a more generalised notion of $\leftrightharpoons$, as expressed in Definition 21. Instead of requiring inputs to be defined for all states reachable by *Utraces*, Definition 21 only requires inputs to be defined for one specific trace.

Lemma 8. *Let* $i \in \mathcal{IOTS}$, $t \in TTS$, $\sigma \in L_{t \restriction i}^*$, $q_i, q_i' \in Q_i$, $q_t, q_t' \in Q_t$:

$$
q_t \text{ }]\!| \text{ } q_i \xrightarrow{\sigma} q_t' \text{ }]\!| \text{ } q_i' \iff q_t \xrightarrow{\sigma \restriction L_t} q_t' \wedge q_i \xrightarrow{\sigma[\theta \mapsto \delta] \restriction L_i^\delta} q_i'
$$

Definition 21. *Let* $\sigma \in L_{s||e}^{\delta *}$, $q_s \in Q_s$, $q_e \in Q_e$. *Then* q_s **accepts** q_e **over** σ $(q_s \xleftarrow{\sigma} q_e)$ *iff:*

$$
\forall \sigma', \sigma'' \in L_{i||e}^{\delta *}, \ell \in L_{i||e}^\delta, q_s' \in Q_s, q_e' \in Q_e :
$$
$$
\sigma' \cdot \ell \cdot \sigma'' = \sigma \wedge q_s \| q_e \xrightarrow{\sigma'} q_s' \| q_e' \implies \boldsymbol{out}(q_e') \cap I_s \subseteq \boldsymbol{in}(q_s') \cap U_e
$$

Lemma 9. *let* $q_s, q_s' \in Q_s$, $q_e, q_e' \in Q_e$, $\sigma \in L_{s||e}^{\delta *}$.

$$
q_s \xleftarrow{\sigma} q_e \implies (\text{ } q_s \| q_e \xrightarrow{\sigma} q_s' \| q_e' \iff q_s \xrightarrow{\sigma \restriction L_s^\delta} q_s' \wedge q_e \xrightarrow{\sigma \restriction L_e^\delta} q_e' \text{ })
$$

Function GenEcoTest($X_e : \mathcal{P}(Q_e)$)

Data: $e \in \mathcal{LTS}$
Data: Inputs I_i and outputs U_i of the SUT
Output: A partial test case for i **eco** e, assuming e could be in any state $q \in X_e$

non-deterministically execute one of the following four cases:
(A) Stop testing:
1:     **return** *pass*
(B) Emit a response to the SUT:
2:     choose $\ell \in (\mathbf{out}(X_e) \cup \mathbf{in}(X_e)) \cap I_i$
3:     **return**
4:        ℓ ; GenEcoTest(X_e *after* ℓ) ##
5:        GenInputResponse(X_e)
(C) Observe quiescence from the SUT (Only if $\delta \in out(X_e)$):
6:     **return**
7:        θ ; GenEcoTest(X_e *after* δ) ##
8:        GenInputResponse(X_e)
(D) Simulate internal behaviour in environment
9:     choose $\ell \in (\mathbf{out}(X_e) \cup \mathbf{in}(X_e)) \setminus L_i$
10:    **return**
11:       ℓ; GenEcoTest(X_e *after* ℓ) ##
12:       GenInputResponse(X_e)
(E) Generate internal behaviour of the SUT
13:    choose $\ell \in I_i \setminus L_e$
14:    **return**
15:       ℓ; GenEcoTest(X_e) ##
16:       GenInputResponse(X_e)

Function GenInputResponse($X_e : \mathcal{P}(Q_e)$)

Data: $e \in \mathcal{LTS}$
Data: Inputs I_i and outputs U_i of the SUT
Output: A partial test case describing how to respond to an input from the *SUT*
1:   **return**
3:     $\Sigma \{\ell ; \mathbf{fail} \mid \ell \in (U_i \cap I_e) \setminus \mathbf{in}(X_e)\}$ ##
5:     $\Sigma \{\ell ; \text{GenEcoTest}(X_e \text{ } \textbf{after } \ell) \mid \ell \in U_i \cap \mathbf{in}(X_e)\}$

Lemma 10. *Let $i \in \mathcal{IOTS}$, $\sigma \in Utraces(i \parallel e)$, $q_i \in Q_i$, $q_e \in Q_e$:*

$$i \parallel e \overset{\sigma}{\Rightarrow} q_i \parallel q_e \wedge e \overset{\sigma}{\nrightarrow} i \implies$$
$$\exists t \in \mathcal{TTSL}_i, e, q_t \in Q_t \setminus \{\boldsymbol{pass}, \boldsymbol{fail}\} :$$
$$t \rceil\!\rceil i \xrightarrow{\sigma[\delta \mapsto \theta]} q_t \rceil\!\rceil q_i$$

Theorem 6. *The test suite obtained from all test cases generated by Algorithm 3 is* **complete**.

Algorithm 3: eco test case generation

Input: A specification of the environment $e \in \mathcal{LTS}$
Data: Inputs I_i and outputs U_i of the SUT
Output: A test case for i **eco** e
return
    test case t with $I_t = (U_i \cap I_e) \cup \{\theta\}$, $U_t = U_e \cup (I_e \setminus U_i) \cup (I_i \setminus L_e)$.
    Q_t and T_t are given by GenEcoTest(e *after* ϵ);

---

6   Related Work

The work in this paper is based on earlier work done on compositional model based testing developed in [9], which introduced environmental conformance (**eco**). The original version of **eco** from Frantzen *et al.* is very similar to the theory of mutual acceptance, but with some key differences:

1. **eco** is defined as *the* notion of correctness, without a link to existing conformance relations such as **uioco**.
2. **eco** only requires the upper interface of each component. Acceptable outputs are inferred from inputs of the components, which is not sufficient information to infer **uioco** correctness. Additionally, **eco** assumes that the two components form a closed system, where all observable behaviour is their communication. This limits the applicability of the theory.

In this paper we combine the strong and weak points of mutual acceptance and environmental conformance into a single theory. We lift **eco** to also deal with more complex label sets containing non-synchronised observable behaviour, and drop the requirement that one model must be input enabled. We show that this expanded definition of **eco** is equivalent to mutual acceptance, and give algorithms to test **eco** conformance on-the-fly.

Interface automata [4–6] model both the behaviour of a component and the constraints it puts on its environment. They introduce the notion of friendly vs unfriendly environments where in the former you assume the environment will try to avoid problems. This could likely be combined with our approach to better differentiate between the final composition, which should always be correct, and intermediate compositions which do not necessarily need to be.

Component based design [11] is a popular approach to designing software systems. This naturally leads to also designing tests in a component based way [1,7,10,12]. A challenge here is to relate the coverage and guarantees of the component based tests to those offered by system level tests, to reason about when to stop testing or where to develop more tests. The theory developed in this paper can help to make this connection, by pointing out where the connections between components are not covered by the component level tests.

7   Future Work

Algorithms 2 and 3 look very similar to the algorithm for testing the **uioco** conformance between an implementation and its specification. They are so similar in fact, that you could run both at the same time on the same implementation with very little overhead. This can cut back on testing time, because waiting for the SUT to give responses is often one of the more time consuming parts of testing. Using the same responses for both algorithms can therefore save time. In their current form however, there is still some overhead. The **eco** algorithm chooses inputs based on what the environment can do, while the **uioco** algorithm chooses inputs based on what the implementation must be able to handle. This means that while there is overlap between the two algorithms, running both of them requires using strictly more inputs than either of them on their own. It remains an open question whether doing this extra work results in better quality of testing.

8 Conclusion

We have shown that after extending the environmental conformance relation (**eco**) to deal with the same scope of models as the mutually accepting relation ($\leftrightharpoons$), the two relations are the same. This allows using them interchangeably, using the simpler to reason about trace based $\leftrightharpoons$ in lemmas and proofs to develop theory, and the algorithms developed for the state based **eco**-*bisimulation* to actually confirm if two specifications are mutually accepting. This checking can be done directly on the specifications, or using testing on a black-box implementation. Both approaches allow compositional testing: correctness of all components implies correctness of the system. This means testing can focus on the component level, giving several advantages: components have smaller models than the whole system, making them easier to create, understand and maintain; retesting on replacing components becomes easier; and smaller models lead to smaller counterexamples which makes using the results of failing tests easier. This should lower the barrier to entry for model based testing, as now a couple of (or even just one) component models can get one started, while allowing the possibility to expand further later on if desired, without losing completeness or having to throw away models.

References

1. Beydeda, S., Gruhn, V.: State of the art in testing components. In: Third International Conference on Quality Software, Proceedings, pp. 146–153 (2003). https://doi.org/10.1109/QSIC.2003.1319097. https://ieeexplore.ieee.org/abstract/document/1319097. Accessed 07 Feb 2024
2. Bolognesi, T., Brinksma, E.: Introduction to the ISO specification language LOTOS. Comput. Netw. ISDN Syst. **14**(1), 25–59 (1987). https://doi.org/10.1016/0169-7552(87)90085-7. https://linkinghub.elsevier.com/retrieve/pii/0169755287900857. Accessed 21 Feb 2023
3. van Cuyck, G., van Arragon, L., Tretmans, J.: Testing Compositionality (2024). arXiv: 2407.05028. https://arxiv.org/abs/2407.05028
4. de Alfaro, L., Henzinger, T.A.: Interface automata. ACM SIGSOFT Softw. Eng. Notes **26**(5), 109–120 (2001). https://doi.org/10.1145/503271.503226. https://dl.acm.org/doi/10.1145/503271.503226. Accessed 29 Oct 2021
5. de Alfaro, L., Henzinger, T.A.: Interface theories for component-based design. In: Henzinger, T.A., Kirsch, C.M. (eds.) EMSOFT 2001. LNCS, vol. 2211, pp. 148–165. Springer, Heidelberg (2001). https://doi.org/10.1007/3-540-45449-7_11
6. de Alfaro, L., Henzinger, T.A.: Interface-based design. In: Broy, M., Grünbauer, J., Harel, D., Hoare, T. (eds.) Engineering Theories of Software Intensive Systems. NSS, vol. 195, pp. 83–104. Springer, Dordrecht (2005). https://doi.org/10.1007/1-4020-3532-2_3

7. Deussen, P., Din, G., Schieferdecker, I.: An on-line test platform for component-based systems. In: 27th Annual NASA Goddard/IEEE Software Engineering Workshop, Proceedings, pp. 96–103 (2002). https://doi.org/10.1109/SEW.2002. 1199455. https://ieeexplore.ieee.org/abstract/document/1199455. Accessed 07 Feb 2024

8. Frantzen, L.: Symbolic Conformance (2024). https://doi.org/10.13140/RG.2.2. 12307.69927. https://rgdoi.net/10.13140/RG.2.2.12307.69927. Accessed 30 Apr 2024

9. Frantzen, L., Tretmans, J.: Model-based testing of environmental conformance of components. In: de Boer, F.S., Bonsangue, M.M., Graf, S., de Roever, W.-P. (eds.) FMCO 2006. LNCS, vol. 4709, pp. 1–25. Springer, Heidelberg (2007). https://doi. org/10.1007/978-3-540-74792-5_1

10. Kuliamin, V.V.: Component architecture of model-based testing environment. Program. Comput. Softw. **36**(5), 289–305 (2010). https://doi.org/10.1134/ S036176881005004X

11. Mahapatro, P.K., Padhy, N.: Reviewing the landscape: component-based software engineering practices and challenges. In: 2024 International Conference on Emerging Systems and Intelligent Computing (ESIC), pp. 360–365 (2024). https:// doi.org/10.1109/ESIC60604.2024.10481576. https://ieeexplore.ieee.org/abstract/ document/10481576. Accessed 28 May 2024

12. Schätz, B., Pfaller, C.: Integrating component tests to system tests. Electron. Notes Theor. Comput. Sci. **260**, 225–241 (2010). https://doi. org/10.1016/j.entcs.2009.12.040. https://www.sciencedirect.com/science/article/ pii/S1571066109005222. Accessed 07 Feb 2024

13. Tretmans, G.J.: Test generation with inputs, outputs and repetitive quiescence. Softw. Concepts Tools **17**(3), 103–120 (1996). https://research.utwente.nl/en/ publications/test-generation-with-inputs-outputs-and-repetitive-quiescence-2. Accessed 09 Nov 2021

14. van Cuyck, G., van Arragon, L., Tretmans, J.: Compositionality in model-based testing. In: Bonfanti, S., Gargantini, A., Salvaneschi, P. (eds.) ICTSS 2023. LNCS, vol. 14131, pp. 202–218. Springer, Cham (2023). https://doi.org/10.1007/978-3-031-43240-8_13

15. van der Bijl, M., Rensink, A., Tretmans, J.: Compositional testing with IOCO. In: Petrenko, A., Ulrich, A. (eds.) FATES 2003. LNCS, vol. 2931, pp. 86–100. Springer, Heidelberg (2004). https://doi.org/10.1007/978-3-540-24617-6_7

Formal Models

Correct Pattern-Based Development Through Refinements and Weakest Preconditions Calculus

Elie Fares[1,2]([⊠])[iD], Jean-Paul Bodeveix[2,3][iD], and Mamoun Filali[3][iD]

[1] Higher Colleges of Technology, Ras Al Khaimah, UAE
[2] IRIT UPS Université de Toulouse, Toulouse, France
`efares@hct.ac.ae, bodeveix@irit.fr`
[3] IRIT CNRS, Université de Toulouse, Toulouse, France
`filali@irit.fr`

Abstract. This paper focuses on the preliminary steps of developing safety-critical systems. We investigate how patterns could be used to generate Event-B refinements automatically. The patterns proposed in this paper either impose constraints on the model through weakest precondition calculus, superpose counters, or introduce de-synchronization mechanisms using observers. We show how the weakest precondition calculus integrated into the patterns can be simplified and become usable. Moreover, for our patterns to be fully usable in subsequent refinement steps, we produce as much as possible readable Event-B models. Finally, we revisit a classic case study using our proposed patterns.

Keywords: Formal requirements · Event-B · Patterns · Weakest preconditions calculus

1 Introduction

Event-B [1] is a formal method for system-level modeling and verification that is based on incremental model development through refinement. In this context, refinement is a process of transforming an abstract model into a more concrete one with guaranteed conformance through the verification of proof obligations. However, these refinement steps remain challenging. The first challenge is the need to abstract a real-life system into an Event-B model. This by itself needs a certain level of expertise. Additionally, the main challenge is the need for formal proofs of every refinement step. Even though several of these proofs are automated, a large number of them need to be done manually. This tends to be complex and time-consuming.

In this context, pattern-based development [15] in Event-B refers to the repeated use of patterns to create complex systems. This approach aims to

This work is partially supported by the ANR project EBRP (Projet-ANR-19-CE25-0010).

D. Marmsoler and M. Sun (Eds.): FACS 2024, LNCS 15189, pp. 59–78, 2024.
https://doi.org/10.1007/978-3-031-71261-6_4

reduce the complexity of the models, increase their consistency and structure, and enhance their readability for future reuse.

In this paper, we propose to use design patterns to produce refinement steps through dedicated domain-specific languages (DSL) [14]. The proposed design patterns are used to generate refinements of existing models. A challenge of the pattern-based approach is to produce both correct and readable models that can later be used by the developer. The correctness property is specified as a refinement step leading to a set of proof obligations (automatically generated) that remain to be discharged. Moreover, hard-to-read models would make future refinement-based developments more difficult. This is why, readability is a key point of this approach as the developer needs to understand both the patterns and the models generated in the process.

To guarantee the usability of the patterns and the readability of the generated models, we first make sure that the transformation does not introduce identifiers (events, variables, labels ...) that have not been provided by the user. The needed identifiers may be introduced by the user along with the property declared through the DSL. Second, the global structure of the model is preserved. Third, we show how the weakest precondition calculus integrated into our patterns may be simplified.

The paper [11] introduces pattern-based refinement for Event-B. Our paper goes further on the following points. First, we give the semantics framework of our pattern application through a composition-based approach. Second, we emphasize the **weakest precondition** (hereafter denoted as **wp**) calculus in the generated machines and we show how this calculus can be simplified using generic and custom rewrite rules. These simplification rules are automatically applied to the generated source code, rather than during the interactive proof process as is typically done. Third, we extend the refinement patterns with a novel synchronous version of the observation pattern. Together with the asynchronous observation pattern, they will allow to safely break an implicit test and set (atomic) by introducing a separate event that performs the test. Fourth, we fully develop the case study by giving the DSLs for every refinement step. Finally, we also extend the case study with temporal and timed properties using our defined patterns. The paper is organized as follows: Sect. 2 introduces our modeling framework Event B. Section 3 introduces our running example which is the well-known mainland/island bridge controller case study [1]. Section 4 describes our refinement-based pattern approach and explains our proposed patterns. In Sect. 5, we rebuild the discussed case study using our refinement patterns. In Sect. 6, we showcase the main related work before drawing a conclusion and discussing the future work in Sect. 7.

2   A Brief Overview of 4Event-B

Event-B [1] is a modeling approach for specifying systems through atomic **events**. The events represent discrete transitions over the state of the system described by **variables** and **invariants**. Invariants are properties satisfied by the initial state and preserved by each event. The Event-B method is supported by the Rodin [2,28] environment that contains a model editor, a static type-checker, a proof obligations generator, and an interactive proof assistant.

Superposition and Data Refinement. We can consider two kinds of refinement: data refinement (vertical refinement) and superposition refinement (horizontal refinement). In data refinement, the purpose of the concrete machine is to change the representation of the state by replacing certain machine variables. The refined events (using the keyword **refines**) simulate the effects of their parent events using the new set of variables.

In superposition refinement [17,24], state variables of the abstract machine are preserved while new variables may be added. The refined events are defined through the keyword **extends** which means that all parameters guards and actions of the parent event are inherited. New parameters, guards and actions may be introduced in the refined event. Moreover, actions —whether added to existing events or belonging to new events— can only modify new variables. Thus, the refinement relation between an event and its extension is verified by construction. More generally, if all events are either new or defined through the **extends** keyword and do not modify inherited variables, then the whole concrete machine is a refinement of the abstract machine by construction.

Proofs. Proof obligations for well-formedness (correctness of partial function application), invariant preservation, refinements and theorems are automatically generated by Rodin. They can be discharged using automatic proof engines (SMT-based [9]: CVC4, Z3, veriT . . .), built-in (PP, AP . . .) or through assisted proofs. These proofs assert that the development has been validated.

In Event-B, most of the proof obligations are based on the *weakest precondition* concept. Given a state predicate P, and an action A, $[A](P)$ is the weakest state predicate defining the states from where the execution of A can only lead to states satisfying P. Weakest preconditions are used to compute the proof obligations associated to the invariants which hold initially and are preserved after the execution of each event.

As stated before, proof obligations are generated automatically and must be discharged either automatically or in an assisted way.

Event-B Machines Composition. The composition of Event-B machines [32,33] is inspired by the CSP parallel operator [22]. A new machine is defined as the synchronous product of several existing machines (the composed machines). The product machine is defined by importing the variables and the invariants of the composed machines. It is also possible to add new variables and invariants. We assume that there are no naming conflicts between variables (thanks to prefixing). An event is defined by importing guards and actions of at most one event of each composed machine. Guards and actions can also be directly added. The product machine can also be declared to refine an existing machine. If imported variables are not modified in the product, the product machine refines each component. This is the same as for the case of the **extends** mechanism explained earlier. This refinement property will be used for establishing the correctness of pattern application (see Sect. 4). In our approach, we use the *CamilleX* plugin [20] to compose machines. The plugin introduces two commands to define products of machines: **includes** inserts variables and invariants of a given machine; **synchronises** inside an event inserts parameters, guards and actions of a given event of an included machine (see Sect. 4).

3   Running Example

This system is given in [1]. The system controls cars on a bridge that links the mainland to a small island. The controller is equipped with two traffic lights. One of the traffic lights is situated on the mainland. The other one is on the island. Four car sensors detect the presence of cars. These sensors are situated at both ends of the bridge (Fig. 1). This system has two main constraints: the number of cars on the island-bridge pair is limited, and the bridge is one-way.

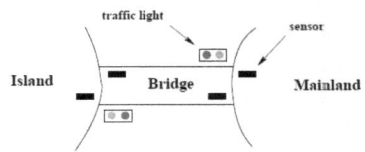

Fig. 1. Abrial's Bridge controller [1]

3.1   Existing Approach

The solution proposed in [1] follows a 3-step refinement strategy. In the initial model, the traffic lights, and the sensors are simply ignored. The bridge is supposed to be part of the island and forms the island-bridge pair. The only constraint considered at this level is limiting the number of cars on this pair. The initial model is represented in the following machine in which two events are modeled. ML2BR represents a car leaving the MainLand and entering the BRidge. This increments the number of cars on the island-bridge. BR2ML represents a car leaving the BRidge and re-entering the MainLand. The latter decreases the number of cars on the island-bridge. The theorem[1] DLF (for deadlock-freedom) expresses the fact that it is always possible to fire one of the events.

```
context cCapacity
constants capacity
axioms capacity ∈ ℕ
end
```

```
machine mac1 sees cCapacity
variables  cars // cars on the bridge
invariants
   @cars_ty cars ∈ 0 .. capacity //closed interval
   theorem @DLF cars<capacity ∨ 0 <cars // disjunction of event guards
events
event INITIALISATION then cars:=0 end
event ML2BR where cars<capacity then cars:=cars+1end //mainland to bridge
event BR2ML where 0<cars then cars:=cars−1 end  // bridge to mainland
end
```

The first refinement introduces the unidirectional bridge which is now decoupled from the island. The cars variable is split into 3 variables, entering cars, isle

---

[1] It is not an axiom. The resulting proof obligation has been discharged.

cars, and exiting cars. Two additional events are also added. BR2IL represents a car leaving the BRidge and entering the IsLand. This increments the isle cars and decrements the entering cars. IL2BR represents a car leaving the IsLand and entering the BRidge. This increments the exiting cars and decrements the isle cars.[2]

The second refinement introduces the **traffic lights** which control the entrance to the bridge at both ends. This refinement adds two variables il_tl and ml_tl containing the two colors (red and green) of the two traffic lights. Cars are not supposed to pass on a red traffic light, only on a green one. A green light ensures safe access to the bridge. The last refinement introduces four car sensors each with two states: on or off. They are used to detect the presence of cars entering or leaving the bridge and thus to update the car counters.

Abrial's original development explained here will be reconstructed in Sect. 6 using our proposed patterns.

4 Pattern-Based Refinement

In this section, we explain the proposed approach and illustrate it in an introductory example. We then explain how behavioral constraints can be integrated into the patterns through the use of the weakest preconditions calculus. We finally show how this calculus can be simplified through the use of built-in and custom simplification rules.

To differentiate between the classic Event-B syntax and our extensions, our keywords are styled in italics and purple, in contrast, to bold for the Event-B keywords.

4.1 Proposed Approach

Figure 2 depicts the outline of the approach. To apply a pattern, a user starts by using a DSL on an initial Event-B machine M_i. The tool generates a pattern instance, composes this pattern instance with the initial Event-B machine, and finally

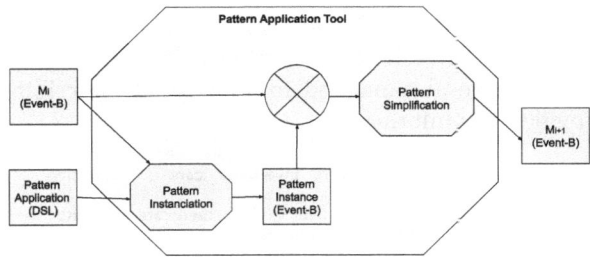

Fig. 2. Pattern application

simplifies the composition result to reach a new Event-B machine M_{i+1} that refines both M_i and the used pattern. In this section, we also give translation-based semantics to our patterns. The semantics of a pattern is a parameterized

---

[2] At the time of this case study writing, lexicographic variants were not yet available in Rodin.

Event-B machine. A pattern application builds a refinement of the machine on which it is applied. By construction, the resulting machine refines an instance of the pattern.

Semantics. Each pattern is represented by a dedicated Event-B machine that sees a context containing the pattern parameters. This machine will denote the pattern semantics. As an example, we introduce the addition of a counter to a machine. The following counter machine gives the semantics of a counter through its unique event INC.

```
machine counter sees counterParam
  variables cnt invariants cnt ∈ Domain
events
  event INITIALISATION then cnt := 0 end
  event INC when cnt+1 ∈ Domain then cnt := cnt + 1 end
end
```

```
context counterParam
constants Domain
axioms
  @Dom Domain ⊆ ℤ
  @D0 0 ∈ Domain
end
```

Pattern Application Specification. A pattern is applied to a source machine (M0) parameterized by pattern arguments. These arguments are described using a DSL in which we give the name of the concerned events (ev1), variables (cars), and possibly predicates (the constraint) or expressions (the expression). The DSL is implemented by an XText plugin [5] which generates an Eclipse editor. As an example, we give the machine M0 and the DSL that applies the counter pattern to this machine. It provides the names of the counters to be added, their domains through constraints, and the events that increment the counter.

```
machine M0
events
  event ev1 end
end
```

```
counting refinement M1 refines M0 sees cCapacity
  counters cars incremented by ev1
  constraints cars ∈ 0..capacity
end
```

```
context cCapacity
constants capacity
axioms
  @capacity∈ capacity ∈ ℕ
end
```

Pattern Application Result. The pattern application will create a machine that refines both, the user's machine and an instance of the pattern semantics. Keeping our same example, we derive M1_counterInstance by renaming the counter variable cnt by cars and selecting an interval 0..capacity for Domain. This operation can be automatized from the DSL describing the pattern instance that provides this information.

```
machine M1_counterInstance sees cCapacity
  variables cars invariants cars ∈ 0..capacity
events
  event INITIALISATION then cars := 0 end
  event INC when cars+1 ∈ 0..capacity then cars := cars + 1 end
end
```

The application of the counter pattern on the machine M0 will produce the machine M1. This latter is declared to refine M0 (proof obligations are generated for this purpose) and is defined as a product of M0 and the pattern instance M1_counterInstance. The *CamilleX* keyword **includes** adds the variable cars and the invariant cars ∈ 0..capacity of M1_counterInstance to M1. The **synchronises** keyword builds the product of event ev1 of M1 and event INC of M1_counterInstance. Thus, M1 also refines the pattern instance by construction. M1_result is the semantics of M1 as defined by *CamilleX*.

machine M0
events
    event ev1
  end
end
	machine M1 refines M0
	includes M1_counterInstance
	events
	    event INITIALISATION
	        synchronises INITIALISATION end
	    event ev1 synchronises INC end
	end
		machine M1_result refines M0
		variables cars invariants cars ∈ 0..capacity
		events
		    event INITIALISATION then cars := 0 end
		    event ev1 when cars+1 ∈ 0..capacity
		        then cars := cars+1 end
		end

4.2   Constrained Behaviors and Weakest Precondition Calculus

The refinement-based development process incrementally introduces behavioral constraints to ensure that the system satisfies given properties [1]. For this, our studied patterns follow this direction by adding these constraints as invariants and strengthening event guards to guarantee their preservation by each event. This strengthening is done by calculating the **wp** of the event's action and the behavioral constraint. Consider the general non-deterministic assignment: $x_1, \ldots, x_n :| BA(x_1, \ldots, x_n, x'_1, \ldots, x'_n)$ where BA is a predicate over old and new variables. Applying the **wp** calculus formula on this assignment and a post-condition Q gives the following formula:

$$[x_1, \ldots x_n :| BA](Q) \equiv \forall x'_1 \ldots x'_n, BA \Rightarrow Q[x_1 := x'_1, \ldots, x_n := x'_n] \qquad (1)$$

In the case of deterministic assignment, the formula can be simplified as shown in the explanatory example of Sect. 4.1 that produces the guard of the event INC for the constraint cars ∈ 0..capacity used as the post-condition. This calculus is automatically performed by our tool.

$$[\text{cnt} :=\text{cnt}+1]\ (\text{cnt} \in 0..\text{capacity}) \equiv \text{cnt} + 1 \in 0..\text{capacity}$$

Using guards strengthening and **wp** calculus provides two advantages. First, our patterns will alleviate the complexity of the manual adding of guards. Second, our guard strengthening will be accurate. This means that our calculated guards are neither too strong thus turning the model overly refined, nor too weak thus leading to failed proofs.

4.3   Formula Simplification

The formulas obtained by the **wp** calculus tend to become rather complex. For this reason, our tool uses the Rodin internal API to call expressions and formula simplifiers on the generated Event-B model. Actions are also simplified (not supported by Rodin). These simplifications are done either by using built-in Rodin rewrite rules [29] or custom rewrite rules. Moreover, the user can safely extend this rule base provided these rules are proved correct. This extension is supported by the Rodin plugin Theory [21]. The rules defined using this plugin are then appropriately applied to simplify the generated model.

Before introducing the simplification rules, let us note that Boolean expressions are not propositions. The Boolean type only contains the constants TRUE and FALSE. There are no operators on this enumerated type. Propositions such as $x < y \wedge ...$, ⊤ (representing truth), and ⊥ (representing falsity) are not Booleans but can be converted to Booleans via the bool function. For example, the conjunction of two Booleans b_1 and b_2 will be written bool(b_1=TRUE $\wedge$ b_2=TRUE).

Built-In Rules. The Rodin rewriting rules handle propositional logic, first-order logic, arithmetic, and set theory. For instance $\top$ can be eliminated from a conjunction thanks to the SIMP_SPECIAL_AND_BTRUE rule $P \wedge \ldots \wedge \top \wedge \ldots \wedge Q \triangleq P \wedge \ldots \wedge Q$.

Custom Rules Using the Theory Plugin. In our case study, we have added the following rewrite rules to the Rodin rules. To the best of our knowledge, these rules are not yet implemented. In the following, conditional rewriting rules are denoted as $p \Rightarrow l \triangleright r$ which is read as if the proposition p (optional) is satisfied then the expression or proposition l is rewritten as the expression or proposition r provided that l and r are both expressions of the same type or propositions. We remark that the rules of the last two rows are conditional. Hypotheses are

bool($b = $ TRUE) $\triangleright b$	$\{$TRUE $\mapsto$ FALSE, FALSE $\mapsto b\}(b) \triangleright$ FALSE
$x \in u..v \Rightarrow x + 1 \in u..v \triangleright x < v$	$x \in u..v \Rightarrow x - 1 \in u..v \triangleright x > u$
$x \geq 0 \Rightarrow x + 1 = 0 \triangleright \bot$	$x \geq 0 \Rightarrow x + 1 \geq 0 \triangleright \top$

looked for in axioms and machine invariants. Moreover, the notation $\{$TRUE $\mapsto$ FALSE, FALSE $\mapsto$ b$\}$ defines a function mapping TRUE to FALSE and FALSE to b.

Simplification Examples. We show some of the simplifications that were automatically done during the application of the Asynchronous Observation Pattern with protection (paragraph 5.3). The two examples depict an assignment using a conditional expression and the result of a **wp** calculus performed to guarantee the constraints specified by the Asynchronous Observation Pattern.
The first example obtained by applying the **wp** calculus is the following assignment:

```
il_out_10 := { TRUE ↦ FALSE, FALSE ↦ il_out_10} (bool(il_out_10=TRUE ∧ ¬(red = green)))
```

The following rules are applied in sequence: (1) Formulas present in axioms or invariants are replaced by $\top$. This is the case for $\neg$(red = green) which results from the specification of the Color enumerated type. (2) $P \wedge \top \triangleright P$. (3) bool($x = $ TRUE) $\triangleright x$. This is a custom rule. (4) $\{$TRUE $\mapsto$ FALSE, FALSE $\mapsto x\}(x) \triangleright$ FALSE. This is a custom rule.

Applying these built-in or custom rules on the right-hand side of the assignment leads to ml_out_10 := FALSE.
The second example obtained by **wp** calculus is the following assignment:

```
ml_out_10 := { TRUE ↦ FALSE, FALSE ↦ ml_out_10} (bool(ml_out_10=TRUE ∧ not(green = green)))
```

Its right-hand side is rather complex but the simplification rules reduce it, thus allowing the elimination of the action. The following conditional rewrite rules are applied in sequence: (1) $x = x \triangleright \top$ (applied with meta variable x = green). (2) $\neg\top \triangleright \bot$. (3) $P \wedge \bot \triangleright \bot$. (4) $\mathrm{bool}(\bot) \triangleright \mathrm{FALSE}$. (5) $\{\mathrm{TRUE} \mapsto x, \mathrm{FALSE} \mapsto y\}(\mathrm{FALSE}) \triangleright y$. Applying these built-in rules on the right-hand side of the assignment leads to il_out_10 := ml_out_10. Our tool recognizes this useless assignment and does not generate it.

5   Proposed Patterns

In this section, we propose the resource introduction pattern, the counter introduction pattern, and the observer-based patterns (synchronous and asynchronous). All of our proposed patterns may be coupled with constraint declarations that will restrict the allowed behavior of the system.

5.1   The Resource Pattern

In this resource synchronization pattern, two events communicate using a bounded number of resources. One event allocates a resource while the other releases it.

Semantics. The semantics is defined by the following machine in which the events are guarded by constraints on the shared resource availability. The event alloc is enabled if a free resource is available. The free resource becomes allocated. The event free is enabled if the provided parameter is allocated. The parameter is then deallocated.

```
context cResource sets  Resource end
machine mResource sees cResource
  variables  allocated        invariants  allocated ⊆ Resource
  events
    event INITIALISATION then allocated := ∅ end
    event alloc any rsc where rsc ∉ allocated then allocated := allocated ∪ {rsc} end
    event free any rsc where rsc ∈ allocated then allocated := allocated \ {rsc} end
end
```

Pattern Application. The resource pattern application is driven by a specific DSL introducing resource sets and resources allocated and released by user events. The syntax of the DSL is not given but is illustrated by the following example. We define the machine M1 as the refinement of a machine M0 that uses resources through its events ev1 and ev2.

```
resource refinement M1 refines M0 sees cTicket // Ticket declaration
  using new resourceSet tickets ⊆ Ticket // renames the allocated variable
  and new resource rsc ∈ Ticket // from context cTicket
    event ev1 allocates resource rsc // ev1 plays the role of alloc
    event ev2 releases resource rsc // ev2 plays the role of free
end
```

Pattern Instance. The refinement schema leads to the generation of the resource instance machine where Resource has been replaced by Ticket and allocated by tickets.

```
machine mResourceInstance sees cTicket
  variables tickets    invariants tickets ⊆ Ticket //from resourceSet declaration
  events
    event INITIALISATION then tickets := ∅ end
    event alloc any rsc where rsc ∉ tickets then tickets := tickets ∪ {rsc} end
    event free any rsc where rsc ∈ tickets then tickets := tickets \ {rsc} end
  end
```

Refined Machine. Applying the resource pattern will produce the following machine M1 which is defined as the composition of M0 and mResourceInstance.

```
machine M1 refines M0 includes mResourceInstance
  events
    event INITIALISATION extends INITIALISATION synchronises INITIALISATION end
    event ev1 extends ev1 synchronises alloc end    event ev2 extends ev2 synchronises free end
  end
```

5.2 Introducing Counters

Event counters are a technique used to keep track of the number of occurrences of a specific event or set of events. Counters may either be general or dedicated. Since we are interested in reasoning over the occurrence of events, we introduce event counters.

Event Counters. Event counters are incremented and decremented by given sets of events. They allow the specification of event-based properties such as precedence properties, producer/consumer properties, bounded drift, ETC. The proposed DSL allows the explicit enumeration of events incrementing or decrementing the counters.

Semantics. The semantics of this pattern is given in the following machine. The action of INC (DEC) increases (decreases) the event counter.

```
machine mCounter
  variables counter    invariants counter ∈ ℤ
  events
    event INITIALISATION then counter := 0 end
    event INC then counter := counter + 1 end    event DEC then counter := counter − 1 end
  end
```

Counting DSL Syntax. The DSL of the counting pattern application is described by the following grammar. We illustrate some features of this DSL in the following sections.

```
CountingRefinement:
  'counting' 'refinement' IDENT
  (' refines ' MACHINE_ID)?
  ('sees' CONTEXT_ID+)?
  ('parameters' EventParameter+)?
  ('events' NewEvent+)? ('counters' Counter+)
  (' invariants ' NamedPredicate+)?
  (' constraints ' NamedPredicate+)?
  (' variants ' Variant+)?
  'end';
```

```
Variant : (' @' IDENT)? expression;
EventParameter: EVENT_ID
  '(' Param (',' Param)* ')'·
  'where' NamedPredicate+;
NewEvent : IDENT
  ('(' Param(',' Param)* ')''where' Guard+)?;
Counter : IDENT
  ('('Param (',' Param)* ')' 'where' Guard+)?
  ('incremented' 'by' Trigger(',' Trigger )*)?
  ('decremented' 'by' Trigger(',' Trigger )*)?;
```

Pattern Application. The counting pattern application is driven by a specific DSL which is used to generate a refinement machine M1 of

```
counting refinement M1 refines M0 sees ctx
counters cnt incremented by evt1 decremented by evt2
constraints @inv cnt ∈ 0..M
end
```

M0 and a counter cnt incremented by evt1 and decremented by evt2 (evt1 and evt2 being declared in M0).

Pattern Instance. The mCounterInstance machine is obtained from mCounter by renaming cnt and introducing the constraint. The **wp** calculus of the events and the constraints

```
machine mCounterInstance sees ctx
  variables cnt              invariants @inv cnt ∈ 0..M
  events
    event INITIALISATION then cnt := 0 end
    event INC when cnt < M then cnt := cnt + 1 end
    event DEC when cnt > 0 then cnt := cnt − 1 end
  end
```

leads to the addition of event guards which were simplified using the rules $x \in u..v \Rightarrow x + 1 \in u..v \triangleright x < v$ and $x \in u..v \Rightarrow x - 1 \in u..v \triangleright x > u$ of § 4.3.

Refined Machine. Applying the counter pattern will produce the following machine M1 which is defined as the composition of M0 and mCounterInstance.

```
machine M1 refines M0 includes mCounterInstance
events
  event INITIALISATION synchronises INITIALISATION end
  event evt1 extends evt1 synchronises INC end       event evt2 extends evt2 synchronises DEC end
end
```

5.3   Observation Patterns

We propose to introduce a refinement by separating, for a given event, the evaluation of its guard from the computation of its action. It follows that guard evaluation and action computation are no longer necessarily atomic. Generally, this pattern aims to replace complex guards that access internal variables invisible to a subsystem (the environment for example) by a test of an observable state variable (a traffic light for example).

For this purpose, we introduce a control variable along with an additional detection event. The control variable is updated by certain events and observed by the target event. As for the detection event, it computes the guards' values and updates the control variable correspondingly. Thus, the testing of complex guards is reduced to testing the value of a control variable (the guards are removed and replaced by the control variable). It is ensured that when the control variable's value is ok, the removed guards are satisfied. This property is thus added as an invariant.

We introduce two variants for this pattern: a synchronous variant where the control variable is computed each time the observed guards are changed

```
machine M0
events
  event evt when G ∧ Gr then A end
  event other when G_other then A_other end
end
```

and an asynchronous variant where the synchronous update is broken by adding an auxiliary event to reset the control variable. In both variants, the refinement patterns are applied to the machine M0 where the guard G^3 of the target event evt will be synchronously/asynchronously observed while Gr remains a guard of evt.

Semantics. The ObsSemantics machine declares an observation variable obs that has two values: ok and ko. The state change is controlled by the refined machine M1.

```
context ObsParams
sets Status
constants ok ko
axioms
    partition ( Status ,{ ok},{ko})
end
```

```
machine ObsSemantics sees ObsParams
variables obs      invariants obs ∈ Status
events
    event INITIALISATION then obs := ko end
    event whenOK when obs = ok end
    event toOK when obs = ko then obs := ok end
    event toKO when obs = ok then obs := ko end
end
```

Observation DSL Syntax. The DSL of the observation pattern application is described by the following grammar that declares **observation** and **exclusion** constraints. An observation is defined by an observer event, a list of observed guards G (**observes** keyword), and the observation variable that may be assigned values that replace ok (**enabled by**) and ko (**disabled by**). The exclusion constraints specify that at most one of the listed observation variables can be enabled. The observed guards are identified by the labels (prefixed by @) used in the refined Event-B model.

```
ObservationRefinement:
('sync'|'async') 'observation' 'refinement' IDENT
(' refines ' MACHINE_ID)?
('sees' CONTEXT_ID+)?
Observation*
Exclusion*
'end';
Exclusion: 'exclusion' VAR_ID+;
```

```
Observation: ' protected '?
'event' EVENT_ID ('sync'|'async')
'observes' ('@' GUARD_ID+)?
'using' 'new' ' variable ' VAR_ID
(':' expression
'enabled' 'by' expression (:= expression )?
'disabled' 'by' expression
)? ('set' 'by' 'new' 'event' EVENT_ID)?;
```

The sync and async variants differ in the way the observation variable is linked to the observed guards. In the synchronous case, the corresponding invariant is $G \Leftrightarrow obs = ok$. Thus, we can replace the guard G of event evt by $obs = ok$. This invariant is preserved by computing the value of obs in each event ev that updates variables of G: obs becomes ok iff G becomes TRUE after executing the event ev. This is again done by calculating the weakest precondition (see Eq. (1)) of the event and the observed guard. Namely, the new value of obs is defined as the result of the application of the function { TRUE ↦ ok, FALSE ↦ ko } which maps TRUE to ok and FALSE to ko to the Boolean value obtained by computing the **wp** of the event evt and the guard G. Our tool will replace [evt](G) by the actual **wp** .

```
obs := { TRUE ↦ ok , FALSE ↦ ko} (bool([evt](G)))
```

In the asynchronous case, the invariant becomes $obs = ok \Rightarrow G$. In the previous sync case, the observation was updated each time the guard was changed. Whereas, in the async case, the observation is set to ok by a separate event introduced by **set by new event** when its guards are satisfied.

---

[3] G could also be a conjunct of guards.

```
event evt_detect when obs = ko ∧ G then obs := ok end
```

The protected keyword disables all events except evt (named other) that falsify the property G. For this purpose, the following guard is added to all such events that modify variables occurring in G. This guard contains a **wp** calculus ensuring that the event execution of the action A_other of the event other will preserve the satisfiability of G.

```
event other extends other where obs = ok ⇒ [other](G) end
```

Asynchronous Observation Pattern. We call it an asynchronous pattern because an auxiliary event enables the control when the guard is satisfied.

Syntax. The following syntax is proposed where g_trigger is the control variable observed by evt, and evt_detect is an auxiliary event that resets the control variable.

```
async observation refinement mac1 refines mac0
   event evt observes guards G1...Gn
   using new variable g_trigger ∈ TYPE enabled by OK disabled by KO
   set by new event evt_detect
end
```

Two variants are introduced: the first variant named Without Protection allows other events to change the value of the control variable. The second variant named With Protection disables all events that would change the value of the control variable.

Without Protection. This variant breaks the atomicity between observation and action. The events that invalidate the observed guards must also update the trigger variable. In the target event, the selected guards are replaced by a test of the trigger variable which is also reset. Other events reset the trigger if the guards are falsified by their actions.

```
machine mac1 refines mac0
 variables g_trigger ... invariants g_trigger=OK ⇒ G1 ∧ ... ∧ Gn
 events ...
     event evt refines evt when g_trigger = OK ∧ Gr then g_trigger := KO
         A
     end
     event other extends other // for all events ≠ evt
     then //if G1 ∧...∧ Gn becomes FALSE, g_trigger is set to FALSE
         g_trigger := {TRUE ↦ g_trigger, FALSE ↦ KO}([A_other](G1 ∧ ... ∧ Gn))
     end
     event evt_detect when g_trigger = KO ∧ G1 ∧ ... ∧ Gn then g_trigger := OK end
 end
```

With Protection. In the following pattern, the atomicity of guards testing and actions is preserved. This is done by introducing a critical section between an event performing the guard's test and the original event performing the action. If the guard succeeds, the subsequent events cannot disable it until the original event occurs.

An event tests the guards and enables another event that can only be triggered once. The next detection will only be allowed after the trigger event has been acknowledged.

The protected keyword precedes the declaration of the target event. The semantics difference between the two variants is that, in all events other than the target event, we replace the trigger update actions with a guard. This is highlighted in the following code using a **wp** calculus ensuring that the execution will preserve the observed guards.

```
event other extends other where g_trigger = OK ⇒ [A_other](G1 ∧...∧ Gn) end
```

Exclusion Constraint. The observation pattern can be used multiple times to introduce several triggers. In such a case, an exclusion constraint can be imposed. Consequently, the evt_guard_detect event associated with a given trigger will reset all other triggers that are in exclusion with it. It follows that the exclusion invariant over each exclusion set is ensured. To enable this feature, we add an **exclusion** clause to the observation patterns that contains a list of control variables. The clause constrains the model so that at any time, at most one control variable is set. The transformation of this constraint consists of two things: (1) An invariant is generated to assert that only one control variable is set at one time. Namely, for a set of control variables $\{c_i \mid i \in 1..N\}$, we generate $\bigwedge_{1 \leq i < j \leq N} c_i = \text{off} \lor c_j = \text{off}$. (2) A guard is added for each event that modifies the control variable. This guard is the **wp** of the event's action and the added invariant. For an illustration example, the reader may refer to Sect. 6.

5.4   Temporal and Timed Patterns

We have already worked in [30,31] on the use of temporal and timed patterns in Event-B. A refinement of the original machine is generated by adding new events such that their execution satisfies the logical property expressed by the pattern. Given a temporal/timed pattern and an Event-B machine, the generation process uses LTL or TPTL [7] logics and timed Büchi automata that are eventually translated to Event-B machines. We do not give the translation approach here but we only illustrate it in the case study section.

6   Development of the Example

We propose a new development chain for the Island-bridge case study. Starting from a model reduced to entering/exiting events, we apply refinement patterns introducing island capacity, car counters and bridge constraints, traffic lights, and lastly car sensors.

```
machine m0_IsleAccess
events
    event ML2BR end // from mainland to bridge
    event BR2ML end // from bridge to mainland
end
```

Initial Model. In our development, we start from a more abstract model than Abrial's one (see Sect. 3.1). We declare two events, one for entering and another for exiting the bridge.

Introduction of Isle-Bridge Capacity. We use the counter pattern to count the vehicles and fulfill the capacity requirement. The definition

```
counting refinement m1_IsleCapacity refines m0_IsleAccess
  sees cCapacity // declares constant capacity
  counters cars incremented by ML2BR decremented by BR2ML
  constraints @capacity cars ∈ 0..capacity
end
```

of this machine will generate implicit guards: $cars < capacity$ for ML2BR and $cars > 0$ for BR2ML.

Introduction of Isle-Bridge Events We use the counter pattern along with new events to count vehicles in each direction of the bridge and on the island.

```
counting refinement m2_BridgeToIsle refines m1_IsleCapacity
events BR2IL IL2BR // new events
counters
  entering incremented by ML2BR decremented by BR2IL
  islecars incremented by BR2IL decremented by IL2BR
  exiting incremented by IL2BR decremented by BR2ML
invariants @glueing cars = entering + islecars + exiting
```

```
constraints
  @exclusivity entering = 0 ∨ exiting = 0
  @enter_ctr entering ≥ 0
  @stay_ctr islecars ≥ 0
  @exit_ctr exiting ≥ 0
end
```

Introduction of Traffic Lights. We use the Asynchronous Observation Pattern Without Protection. Applying this pattern will replace the synchronous test of counters by the events ML2BR and IL2BR through the guards @capacity and @exclusivity. These events will observe the newly introduced traffic lights asynchronously updated by the events ml_green and il_green. The traffic lights play the role of the triggers introduced by the pattern. They are updated each time a vehicle enters or exits the bridge.

```
async observation refinement m3_TrafficLights refines m2_BridgeToIsle
  sees cColor //defines the set Color with constants red and green
  event ML2BR async observes guards @capacity @exclusivity
    using new variable ml_tl∈COLOR enabled by green disabled by red
    set by new event ml_green // ml_tl back to green if guards are T
  event IL2BR async observes guards @exclusivity
    using new variable il_tl ∈COLOR enabled by green disabled by red
    set by new event il_green // il_tl back to green if guard is T
  exclusion il_tl ml_tl // forbids il_tl and ml_tl to be both green·
end
```

Introduction of Car Sensors. We use the Asynchronous Observation Pattern with Protection. We remove the direct access to traffic lights through the guards g_ml_tl and g_il_tl by introducing car sensors. The events ML2BR and IL2BR that compute the new values of the traffic lights will access sensor values instead of directly reading the current traffic lights' values. These events are declared protected since they should not be disabled whenever the green light is observed by the driver.

```
async observation pattern m4_Sensors refines m3_TrafficLights sees cColor
  protected event ML2BR async observes guards @g_ml_tl //guard of ML2BR in m3_TrafficLights
    using new variable ml_out_10 set by new event ML2BR_sensor
  protected event IL2BR async observes guards @g_il_tl //guard of IL2BR in m3_TrafficLights
    using new variable il_out_10 set by new event IL2BR_sensor
  event BR2ML using new variable ml_in_10 set by new event BR2ML_sensor
  event BR2IL using new variable il_in_10 set by new event BR2IL_sensor
end
```

At this point, Abrial's model has been rebuilt by only applying two patterns, each twice.

Generated Model. We show here an excerpt of the event il_green at the car sensors refinement level. The generated event shows two complex actions (among others) that are automatically simplified by our simplification rules. Starting from:

```
event il_green extends il_green then
    @ml_out_10_ko ml_out_10 := { TRUE ↦ FALSE, FALSE ↦ ml_out_10}
      (bool(ml_out_10=TRUE ∧ not(red = green)))
    @il_out_10_ko il_out_10 := { TRUE ↦ FALSE, FALSE ↦ il_out_10}
      (bool(il_out_10=TRUE ∧ not(green = green)))
    ... // 2 more similar actions
```

After simplification, the first action reduces to ml_out_10 := FALSE and the second action il_out_10_ko is reduced to il_out_10 := il_out_10 which is not generated. The same happens for the two remaining actions of this event.

Traffic Lights Fairness. To constrain the system's fairness, we require that the difference between the number of times each traffic light becomes green is bounded. For this, we introduce a counter which is incre-

```
counting refinement m5_Fairness
  refines m4_Sensors sees cColor cCapacity cCategory
  counters ml_il_cpt
    incremented by ml_green
    decremented by il_green
  constraints @fairness ml_il_cpt ∈ −1..1
```

mented by the event ml_green and decremented by the event il_green. These events change the traffic lights to green on both the mainland and the island. The counter's interval is constrained to −1..1 to ensure that events alternate.

Bounded Car Leaving. To illustrate a timed pattern, we consider the environment constraint that specifies that whenever a vehicle leaves the bridge entry sensor, it

```
refinement m6_Leaving refines m5_Fairness
  event IL2BR leadsto event BR2ML
    (using new status leaving)
    within 60 new event Tick (using new clock tick);
```

should reach the exit sensor in less than 60 time units. The generated model for this pattern will block time evolution if BR2ML does not occur within 60 ticks.

7   Related Work

The papers [23, 25] introduce a generic pattern language that allows to produce refinements of Event-B machines. The language provides access to structural elements of Event-B machines (variables, invariants, events, ...) and thus allows adding/removing of such elements from the model. With this, one can define patterns and apply them to Event-B machines. However, their pattern language does not support **wp** calculus.

Concerning patterns, a pioneering work [27] has been developed for Atelier-B. It addresses the automatic generation of a B implementation model from data structures and statements yet to be refined. Additionally, several design patterns have been developed for Event-B. In [18], the authors introduce some classical design patterns in an Event-B development. Patterns are seen as generic Event-B refinement steps. A mapping between the user development and the pattern must be provided to instantiate the generic refinement. In our proposal, we make the mapping explicit through DSL.

In [19], a set of design patterns is studied. These design patterns focus mostly on modeling communications. However, the corresponding Rodin plugin does not capture either expressions or predicates of the user's model. Moreover, the tool does not allow the instantiation of context elements seen as pattern parameters. Our approach allows referencing users' predicates and building new ones. Patterns can be parameterized as the DSL associated with each pattern allows the instantiation of these parameters.

In [3], the focus is on sequencing patterns while [4,8] focus mostly on architectural patterns. In the former, the authors propose a dedicated tool while none is proposed in the last two. Our work is methodologically different: we have not tried to exhaustively cover a certain application domain but we have rather proposed lightweight DSL to ease the pattern application irrespective of the application domain. Concerning model simplification, we can cite rewriting-based external tools such as MAUDE [26], {log} [10]. Some of them (mainly {log}) contain an important number of first-order and set theory reduction rules. However, we have preferred to stick to the internal Rodin simplification mechanism, thus avoiding the back-and-forth transcription of Event-B formulas.

Patterns applications have highlighted the need for high-level operators on Event-B machines and contexts. In our work, this is remedied by the use of *CamilleX* to compose machines and events. However, the renaming of individual variables and the parametrization of machines by contexts are not supported. A proposal for context instantiation is done in the EBRP project [6,16]. A seminal work based on institutions is done in [12,13]. Composition and renaming operators are proposed as basic modularisation constructs. However, they do not allow a smooth manipulation of event guards as we propose. The use of our dedicated patterns (e.g. counters and observations) renders the example lighter since our patterns may be parameterized by event guards. In our opinion, this makes the example more readable and thus more adapted for future reuse.

8   Conclusion

In this paper, we have proposed an approach of Event-B refinements driven by DSL-based patterns. We have extended and detailed the observation/de-synchronization patterns. We have then rebuilt the refinement chain of the case study by only using two patterns. This reverse engineering has led to the exhibit of the concept of constraints which are established as invariants through guards strengthening. These constraints lead to the **wp** calculus to produce the refined model. We have also shown how the **wp** calculus may be simplified to reduce the complexity of the refined model. The correctness of the approach is ensured by discharging (here automatically) the proof obligations associated with the refinement steps. Moreover, we believe that the proposed patterns are usable and the generated machines are human-readable. First, the patterns do not create new identifiers. Second, the patterns preserve the global structure of the model. Third, the patterns do not create complex guards or actions in the refined models since the computed **wp** are simplified using either built-in rewrite rules or custom rewrite rules. It follows that the generated proof obligations are also simplified. More generally, our work highlights the need for a machine simplification library that could be used by any Event-B model generator.

Finally, pattern application is also a way to document and systematize the construction of refinement steps. We use an approach where every refinement step is done through the application of a pattern.

References

1. Abrial, J.-R.: Modeling in Event-B: System and Software Engineering, 1st edn. Cambridge University Press, Cambridge (2010)
2. Abrial, J.-R., Butler, M., Hallerstede, S., Hoang, T.S., Mehta, F., Voisin, L.: Rodin: an open toolset for modelling and reasoning in Event-B. Int. J. Software Tools Technol. Transfer **12**(6), 447–466 (2010)
3. Alkhammash, E., Butler, M., Fathabadi, A.S., Cîrstea, C.: Building traceable Event-B models from requirements. Sci. Comput. Programm. **111**, 318–338 (2015). Special Issue on Automated Verification of Critical Systems (AVoCS 2013)
4. Ball, E., Butler, M.: Event-B patterns for specifying fault-tolerance in multi-agent interaction. In: Butler, M., Jones, C., Romanovsky, A., Troubitsyna, E. (eds.) Methods, Models and Tools for Fault Tolerance. LNCS, vol. 5454, pp. 104–129. Springer, Heidelberg (2009). https://doi.org/10.1007/978-3-642-00867-2_6
5. Bettini, L.: Implementing Domain Specific Languages with Xtext and Xtend - Second Edition, 2nd edn. Packt Publishing (2016)
6. Bodeveix, J.-P., Filali, M.: Event-B formalization of event-B contexts. In: Raschke, A., Méry, D. (eds.) ABZ 2021. LNCS, vol. 12709, pp. 66–80. Springer, Cham (2021). https://doi.org/10.1007/978-3-030-77543-8_5
7. Bouyer, P., Chevalier, F., Markey, N.: On the expressiveness of **TPTL** and **MTL**. In: Sarukkai, S., Sen, S. (eds.) FSTTCS 2005. LNCS, vol. 3821, pp. 432–443. Springer, Heidelberg (2005). https://doi.org/10.1007/11590156_35
8. Bryans, J.W., Wei, W.: Formal analysis of BPMN models using event-B. In: Kowalewski, S., Roveri, M. (eds.) FMICS 2010. LNCS, vol. 6371, pp. 33–49. Springer, Heidelberg (2010). https://doi.org/10.1007/978-3-642-15898-8_3
9. Déharbe, D., Fontaine, P., Guyot, Y., Voisin, L.: Integrating SMT solvers in Rodin. Sci. Comput. Program. **94**, 130–143 (2014)
10. Dovier, A., Piazza, C., Pontelli, E., Rossi, G.: Sets and constraint logic programming. ACM Trans. Program. Lang. Syst. **22**(5), 861–931 (2000)

11. Fares, E., Bodeveix, P.J., Filali, M.: Pattern-based refinement generation through domain specific languages. In: Glässer, U., Creissac Campos, J., Méry, D., Palanque, P. (eds.) ABZ 2023. Pattern-based refinement generation through domain specific languages, vol. 14010, pp. 35–42. Springer, Cham (2023). https://doi.org/10.1007/978-3-031-33163-3_3
12. Farrell, M.: Event-B in the institutional framework: defining a semantics, modularisation constructs and interoperability for a specification language (2017)
13. Farrell, M., Monahan, R., Power, J.F.: Building specifications in the Event-B institution: a summary. In: Glässer, U., Creissac Campos, J., Méry, D., Palanque, P. (eds.) ABZ 2023. LNCS, vol. 14010, pp. 245–253. Springer, Cham (2023). https://doi.org/10.1007/978-3-031-33163-3_19
14. Fowler, M.: Domain-Specific Languages. Addison-Wesley, Upper Saddle River (2010)
15. Gamma, E., Helm, R., Johnson, R., Vlissides, J.M.: Design Patterns: Elements of Reusable Object-Oriented Software, 1st edn. Addison-Wesley Professional (1994)
16. Guillaume Verdier, L.V.: Context instantiation plug-in: a new approach to genericity in Rodin. In: Proceedings of the 9th Rodin User and Developer Workshop (2021)
17. Hoang, T.S.: An introduction to the Event-B modelling method. In: Romanovsky, A., Thomas, M. (eds.) Industrial Deployment of System Engineering Methods, pp. 211–236. Springer, Cham (2013). http://www.springer.com/computer/swe/book/978-3-642-33169-5
18. Hoang, T.S., Fürst, A., Abrial, J.-R.: Event-B patterns and their tool support. In: Hung, D.V., Krishnan, P. (eds.) Seventh IEEE International Conference on Software Engineering and Formal Methods, SEFM 2009, Hanoi, Vietnam, 23–27 November 2009, pp. 210–219. IEEE Computer Society (2009)
19. Hoang, T.S., Fürst, A., Abrial, J.-R.: Event-B patterns and their tool support. Software Syst. Model. 12, 229–244 (2013)
20. Hoang, T.S., Snook, C., Dghaym, D., Fathabadi, A.S., Butler, M.: Building an extensible textual framework for the Rodin platform. In: Masci, P., Bernardeschi, C., Graziani, P., Koddenbrock, M., Palmieri, M. (eds.) SEFM 2022. LNCS, vol. 13765, pp. 132–147. Springer, Heidelberg (2023). https://doi.org/10.1007/978-3-031-26236-4_11
21. Hoang, T.S., Voisin, L., Salehi,A., Butler, M.J., Wilkinson, T., Beauger, N.: Theory plug-in for rodin 3.x. CoRR, abs/1701.08625 (2017)
22. Hoare, C.A.R.: Communicating Sequential Processes. Prentice-Hall International Series in Computer Science. Prentice Hall (1985)
23. Iliasov, A., Troubitsyna, E., Laibinis, L., Romanovsky, A.B.: Patterns for refinement automation. 6286, 70–88 (2009)
24. Kobayashi, T., Ishikawa, F.: Analysis on strategies of superposition refinement of event-B specifications. In: Sun, J., Sun, M. (eds.) ICFEM 2018. LNCS, vol. 11232, pp. 357–372. Springer, Cham (2018). https://doi.org/10.1007/978-3-030-02450-5_21
25. Laibinis, L., Troubitsyna, E., Iliasov, A., Romanovsky, A.: Rigorous Development of Fault-Tolerant Agent Systems, pp. 241–260. Springer, Heidelberg (2006)
26. Ölveczky, P.C., Meseguer, J.: Specifying real-time systems in rewriting logic. In: Meseguer, J. (ed.) Electronic Notes in Theoretical Computer Science, volume 4. Elsevier Science Publishers (2000)
27. Requet, A.: BART: a tool for automatic refinement. In: Börger, E., Butler, M., Bowen, J.P., Boca, P. (eds.) ABZ 2008. LNCS, vol. 5238, pp. 345–345. Springer, Heidelberg (2008). https://doi.org/10.1007/978-3-540-87603-8_33

28. http://www.event-b.org/
29. https://wiki.event-b.org/index.php/Set_Rewrite_Rules
30. Siala, B., Bhiri, M.T.: An automatic refinement for event-B through annotated temporal logic patterns. In: Nguyen, N.T., Manolopoulos, Y., Chbeir, R., Kozierkiewicz, A., Trawinski, B. (eds.) ICCCI 2022. LNCS, vol. 13501, pp. 624–637. Springer, Cham (2022). https://doi.org/10.1007/978-3-031-16014-1_49
31. Siala, B., Bodeveix, J.-P., Filali, M., Bhiri, M.T.: Automatic refinement for Event-B through annotated patterns. In: Kotenko, I.V., Cotronis, Y., Daneshtalab, M. (eds.) 25th Euromicro International Conference on Parallel, Distributed and Network-based Processing, PDP 2017, St. Petersburg, Russia, March 6–8, 2017, pp. 287–290. IEEE Computer Society (2017)
32. Silva, R.: Towards the composition of specifications in Event-B. In: Proceedings of the B 2011 Workshop, a satellite event of the 17th International Symposium on Formal Methods (FM 2011), Electronic Notes in Theoretical Computer Science, vol. 280, pp. 81–93 (2011)
33. Silva, R., Butler, M.: Shared event composition/decomposition in event-B. In: Aichernig, B.K., de Boer, F.S., Bonsangue, M.M. (eds.) FMCO 2010. LNCS, vol. 6957, pp. 122–141. Springer, Heidelberg (2011). https://doi.org/10.1007/978-3-642-25271-6_7

Coq Formalization of Orientation Representation: Matrix, Euler Angles, Axis-Angle and Quaternion

Zhengpu Shi[(✉)] and Gang Chen

Nanjing University of Aeronautics and Astronautics, Nanjing, China
{zhengpushi,gangchensh}@nuaa.edu.cn

Abstract. Rotation matrices, Euler angles, axis-angle, and unit quaternions are common models for representing object pose in space. Each offers distinct advantages and disadvantages regarding handling singularities, computational complexity, and storage requirements, motivating their interchangeability and conversion algorithms. However, these models and their associated algorithms involve complex matrix operations, trigonometric computations, and geometric reasoning, making manual derivation and verification error-prone. To guarantee algorithm correctness, we present a formal verification of these models and their conversion algorithms within the Coq Proof Assistant, focusing on pure rotational orientation representation. We establish a formal foundational mathematical library, including vector and matrix theories on abstract elements, additional real matrix properties, and quaternion theory. Building upon this foundation, we formalize all four rotation models and their conversion algorithms, emphasizing invariants of special orthogonal groups, Rodrigues' rotation formula, and the quaternion rotation formula. Our formally verified library offers excellent readability, usability, and simplicity, laying a foundation for developers to understand and verify advanced kinematics algorithms.

Keywords: Orientation representation · Formal verification · Coq proof assistant · Rotation matrix · Euler angles · Axis-angle · Unit quaternions

1 Introduction

In motion control and computer graphics, accurate representation of object pose (position and orientation) is crucial for system control. For example, aircraft rely on pose information for maneuvering, and computer graphics rendering requires object and camera poses. *Pose representation* ($\mathcal{PR}$) involves two stages: *orientation representation* ($\mathcal{OR}$) with pure rotation, and $\mathcal{PR}$ with both rotation and translation. We focus on $\mathcal{OR}$, an integral and challenging part of $\mathcal{PR}$. Four main $\mathcal{OR}$ models exist, each with advantages and disadvantages: 1. Rotation matrices are composable but lack intuitiveness and require more memory. 2. Euler angles are intuitive but suffer from gimbal lock singularity. 3. Axis-angle is efficient for interpolation but computationally expensive. 4. Unit quaternions offer numerical stability and efficiency but lack intuitiveness. These models are often used together and converted between each other. Other models, such as exponential map and dual quaternions, exist but are not discussed here.

However, manually deriving these models and algorithms is highly complex and error-prone. The derivation involves intricate matrix operations, trigonometric computations, and geometric reasoning, making it easy to introduce human errors, such as incorrect subscripts, symbols, or missing conditions. For example, Euler angles' singularities necessitate special handling when converting from rotation matrices. Research has shown that even derivations by mathematicians may contain errors or flaws [2]. We also identified and confirmed errors in the formulas from a textbook [12]. These errors cause serious safety risks when used in algorithms for safety-critical systems. Therefore, rigorous measures are necessary to ensure the correctness of the derivation process and results.

Verifying $\mathcal{OR}$ algorithms is challenging due to the limitations of traditional verification methods. Exhaustive software testing and model checking are impractical for real-valued inputs, while automated theorem proving struggles with the inherent complexity of $\mathcal{OR}$ algorithms. We leverage Coq [3], a higher-order logic theorem prover, for its dependent type system that facilitates vector modeling, and its expressiveness in formalizing intricate mathematical concepts.

We formalized several $\mathcal{OR}$ models and verified their algorithms using the COQ PROOF ASSISTANT. Our main contributions are:1. A formal mathematical library covering basic vector and matrix theories, as well as quaternion theory. 2. Formalization of complex matrix operations, including determinants, inverse matrices based on Gaussian elimination or adjoint matrices, and the special orthogonal group SO(n, $\mathbb{F}$).3. Formalization of $\mathcal{OR}$ models and related algorithms based on this mathematical library, including:(1) Rotation matrices and multiple invariants;(2) Rotation matrices under 24 conventions of Euler angles, the existence of Euler angles singularities, and conversion from rotation matrices to Euler angles;(3) Axis-angle representation and Rodrigues' rotation formula; (4) Quaternion rotation formula and its conversion with rotation matrices.

The rest of the paper is organized as follows: Sect. 2 presents the formal mathematical library. Section 3 discusses formal verification of $\mathcal{OR}$ algorithms and several key theorems. Section 4 introduces the user interface and practical applications. Section 5 reviews related work. Section 6 concludes the paper.

2   Foundational Mathematical Library

This section presents a formal mathematical library, which fills gaps in the COQ standard library by providing vector and matrix theory, as well as quaternion theory. We formalize various operations and related properties for these mathematical objects. Due to space constraints, only the key concepts used in this paper are provided.

2.1 Generic Vector Theory

This subsection introduces generic vectors over abstract elements. We define vector types and common vector operations, and prove various properties about these operations. Vector indices start from 1. For instance, a 3D vector **a** can be represented as $\mathbf{a} = (a.1, a.2, a.3)$. Refer to Sect. 2.3 for discussions on column and row vectors. We use "%V" as the delimiter for notations of vector types.

Although five formal models of vectors and matrices have been discussed in the literature [14,15], we present a different approach using functions from indices (finite set type) to elements to represent vectors, as defined below:

Definition fin (n : nat) := {i | i < n}. (* Finite set: {0,1,..,(n-1)} *)

Definition vec {A : **Type**} (n : nat) := fin n -> A.

Here, we define the type fin for vector indices and type vec for n-dimensional vectors over elements of type A. The design offers several advantages:1) Type-level checking of vector dimensions prevents mismatched data use.2) Functional definition and proofs simplify vector manipulation compared to structured data types.3) Direct Leibniz equality removes the need for setoid equality. 4) Its simplicity surpasses the definition in MathComp library [10] while allowing expression evaluation. 5) Nested vec supports matrices and multi-level vectors.

Vector element extraction is simply a function application. Conversions between lists and vectors facilitate typing and printing.

Notation "v .1" := (v (@nat2finS _ 0)). **Notation** "v .x" := (v.1).

Definition l2v {A} {Azero:A} n (l:list A) : vec n := λ i ⇒ nth i l Azero.

Definition v2l {A} {n} (v:vec n) : list A := map v (finseq n).

Here, the "@" symbol is used to explicitly provide implicit arguments. @nat2finS n i constructs an object of type fin (S n), and finseq generates a list of type fin n containing all natural numbers less than n.

Table 1 summarizes key vector operations. Parallel and perpendicular components require further explanation, as detailed in Definition 1.

Table 1. Operations & predicates on n-dimensional vectors over abstract element type

Name	Description of function or predicate	Notation(Math)	Notation(Coq)
vadd	vector addition	$\mathbf{a} + \mathbf{b}$	a + b
vcmul	vector scalar multiplication	$k\mathbf{a}$	k \.* a
vdot	vector dot product	$\langle \mathbf{a}, \mathbf{b} \rangle$	<a,b>
vlen	vector length	$\|\mathbf{a}\|$	||a||
vunit	vector **a** is a unit vector?	$\hat{\mathbf{a}}$	
vorth	vectors **a** and **b** are orthogonal?	$\mathbf{a} \perp \mathbf{b}$	a _|_ b

We define vector elements into a hierarchy of types, ranging from semigroups to normed fields. We also construct vector modules over these abstract elements and instantiate them for common number fields (natural numbers, complex numbers, etc.). Details are omitted. We will omit the carrier type A if context allows.

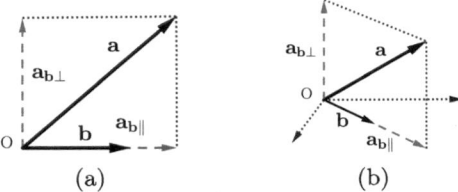

Fig. 1. 2D/3D Examples of parallel/perpendicular components of vector **a** onto **b**

Definition 1 (Parallel and Perpendicular Component). *For any vector* **a** *and non-zero vector* **b***, the projection of* **a** *onto* **b** *can be decomposed into: parallel component* $\mathbf{a_{b\parallel}} \triangleq \frac{\langle \mathbf{a},\mathbf{b}\rangle}{\langle \mathbf{b},\mathbf{b}\rangle}\mathbf{b}$ *and perpendicular component* $\mathbf{a_{b\perp}} \triangleq \mathbf{a} - \mathbf{a_{b\parallel}}$. *See Fig. 1 for illustration, and the formal definitions below.*

```
Definition vproj {n} a b : vec n := (<a,b> / <b,b>) \.* b.
Definition vperp {n} a b : vec n := a - vproj a b.
```

2.2 Vector Theory over Real Numbers

This subsection introduces real vectors (i.e., in Euclidean spaces), where we can define operations such as normalization, angle, and cross product. The operations we have defined for real vectors are shown in Table 2.

Table 2. Extra operations on n-dimensional real vectors

Name	Description of function	Notation(Math)	Notation(Coq)
vnorm	normalize a real vector	$\frac{\mathbf{a}}{\|\mathbf{a}\|}$	–
vangle	angle of two real vectors	$\mathbf{a} \angle \mathbf{b}$	a /_ b
v3cross	3D vector cross product	$\mathbf{a} \times \mathbf{b}$	a \x b
skew3	3D vector to skew-symmetric matrix	$[\mathbf{a}]_\times$	–

The 3D cross product definition below is a natural translation from mathematics to formalization, demonstrating the excellent readability of our library.

```
Definition v3cross (a b : vec 3) : vec 3 := l2v
  [a.2 * b.3 - a.3 * b.2; a.3 * b.1 - a.1 * b.3; a.1 * b.2 - a.2 * b.1].
```

2.3 Generic Matrix Theory

In this subsection, we leverage the previously defined vector theory to formalize general matrix theory. A matrix type is represented as a vector of vectors, essentially a binary function that maps row and column indices to elements.

Definition mat A r c := (@vec (@vec A c) r). **Notation** smat A n :=(mat A n n).

Eval **cbv in** ∀ A r c, mat A r c. (* ∀ A r c, fin r -> fin c -> A *)

Here, mat A r c denotes the $r \times c$ matrix type over A, smat A n is an abbreviation for the $n \times n$ square matrix type over A. We will use a delimiter "%M" for matrix.

When matrices and vectors are used together, distinguishing between column and row vectors is essential, as popular graphics APIs have different conventions: OpenGL uses matrices to multiply column vectors, while DirectX uses row vectors to multiply matrices. We define column vectors as $n \times 1$ matrices and row vectors as $1 \times n$ matrices, using notation $\mathbf{a}$ and $\mathbf{a}^\mathrm{T}$ separately. Formally:

Notation cvec A n := (mat A n 1). **Notation** rvec A n := (mat A 1 n).

We formalize fundamental matrix operations, such as transpose and multiplication, along with advanced functions like determinant and inverse, as listed in Table 3. The theory of orthogonal matrices are detailed in Sect. 3.1.

Table 3. Operations and predicates on matrices over abstract element type

Name	Description of function or predicate	Notation(Math)	Notation(CoQ)
l2m	convert list of list to matrix		
mtrans	matrix transpose	$\mathbf{A}^\mathrm{T}$	A\T
madd	matrix addition	$\mathbf{A} + \mathbf{B}$	A + B
mcmul	matrix scalar multiplication	$k\mathbf{A}$	k \.* A
mmul	matrix multiplication	$\mathbf{AB}$	A * B
mmulv	matrix multiply vector	$\mathbf{Au}$	A *v u
mdet	determinant of a matrix	$|\mathbf{A}|$	|A|
minvAM	matrix inversion via adjoint matrix	$\mathbf{A}^{-1}$	A\-1

2.4 Quaternion

Quaternions are powerful mathematical tools for handling $\mathcal{OR}$ problems, featuring non-singularity and low computational cost. This subsection covers their basic operations, while specific applications are detailed in Sect. 3.4.

Definition 2 (Quaternion). *A quaternion* $\mathbf{q} \in \mathbb{H}$ *is a number of the form* $\mathbf{q} = w + xi + yj + zk$ *that combines four real numbers* w, x, y, z *and three imaginary units* i, j, k, *denoted as* $\mathbf{q} = (w, x, y, z)^\mathrm{T}$ *or* $\mathbf{q} = (w, \mathbf{v})^\mathrm{T}$, *which satisfy:*

$$i^2 = j^2 = k^2 = -1, \quad ij = -ji = k, \quad jk = -kj = i, \quad ki = -ik = j, \quad (1)$$

where w *is the scalar part, and* $\mathbf{v} = (x, y, z)^\mathrm{T}$ *is the imaginary part. We represent a quaternion using a 4-dimensional vector. We also use the functions* q2im, im2q, *and* si2q *to convert quaternions and vectors. Formally:*

Definition quat := vec 4. **Notation** "q .W" := (q.1). (* ... *)

Definition q2im (q : quat) : vec 3 := l2v [q.X; q.Y; q.Z].

Definition im2q (v : vec 3) : quat := l2v [0; q.X; q.Y; q.Z].

Definition si2q (w : R) (v : vec 3) : quat := l2v [w; v.1; v.2; v.3].

For any quaternion $\mathbf{q} = (w, x, y, z)^{\mathrm{T}}$, its magnitude is denoted by $|\mathbf{q}|$ and defined as $\sqrt{w^2 + x^2 + y^2 + z^2}$. The conjugate of $\mathbf{q}$ is denoted as $\overline{\mathbf{q}}$ and defined as $\mathtt{qconj}(\mathbf{q}) \triangleq (w, -x, -y, -z)$. The quaternion multiplication is denoted as $\mathbf{q}_1 \otimes \mathbf{q}_2$ in mathematics, and as q1 $\star$ q2 in Coq, where $\mathbf{q}_i = (w_i, x_i, y_i, z_i)^{\mathrm{T}}$.

$$\mathbf{q}_1 \otimes \mathbf{q}_2 \triangleq \begin{pmatrix} w_1 w_2 - x_1 x_2 - y_1 y_2 - z_1 z_2 \\ w_1 x_2 + x_1 w_2 + y_1 z_2 - z_1 y_2 \\ w_1 y_2 - x_1 z_2 + y_1 w_2 + z_1 x_2 \\ w_1 z_2 + x_1 y_2 - y_1 x_2 + z_1 w_2 \end{pmatrix}. \tag{2}$$

Lemma 1 (Matrix Form of Quaternion Multiplication). *For any two quaternions $\mathbf{q}_1$ and $\mathbf{q}_2$, $\mathbf{q}_1 \otimes \mathbf{q}_2$ equals the following matrix multiplication:*

$$\mathbf{q}_1 \otimes \mathbf{q}_2 = \mathtt{ML}(\mathbf{q}_1)\mathbf{q}_2, \qquad\qquad \mathbf{q}_1 \otimes \mathbf{q}_2 = \mathtt{MR}(\mathbf{q}_2)\mathbf{q}_1, \tag{3}$$

$$\mathtt{ML}(\mathbf{q}) \triangleq \begin{bmatrix} w & -x & -y & -z \\ x & w & -z & y \\ y & z & w & -x \\ z & -y & x & w \end{bmatrix}, \qquad \mathtt{MR}(\mathbf{q}) \triangleq \begin{bmatrix} w & -x & -y & -z \\ x & w & z & -y \\ y & -z & w & x \\ z & y & -x & w \end{bmatrix}, \tag{4}$$

where $\mathtt{ML}(\mathbf{q})$ and $\mathtt{MR}(\mathbf{q})$ denote the left and right matrix of $\mathbf{q}$, respectively.

Proof. The first equation in (3) is formally proven as follows:

Definition qmatL (q : quat) := **let** '(w,x,y,z) := (q.1,q.2,q.3,q.4) **in**
    l2m [[w;-x;-y;-z]; [x;w;-z;y]; [y;z;w;-x]; [z;-y;x;w]].
Lemma qmatL_spec: $\forall$ p q, p * q = (qmatL p) *v q. **Proof.** veq; ra. **Qed.**

Here, qmatL is the conversion function $\mathtt{ML}$. The proof is automated using two customized tactics: veq, converting vector equality into element-wise equalities, and ra, a real number automation tactic that manages various simplifications. The meq tactic is also used for matrix equality in subsequent sections.

3 Orientation Representation

This section discusses models and algorithms for $\mathcal{OR}$. We begin by clarifying fundamental concepts to avoid common mistakes [6].

- *Frame:* Due to spatial relativity, we establish two frames: $\mathcal{S}$ (space-fixed) and $\mathcal{B}$ (body-fixed). Denoting a vector $\mathbf{a}$ in $\mathcal{S}$ or $\mathcal{B}$ as ${}^{\mathcal{S}}\mathbf{a}$ or ${}^{\mathcal{B}}\mathbf{a}$, the transformation from $\mathcal{B}$ to $\mathcal{S}$ (also the relative orientation of $\mathcal{B}$ with respect to $\mathcal{S}$) is ${}^{\mathcal{S}}\mathbf{A}_{\mathcal{B}}$, leading to ${}^{\mathcal{S}}\mathbf{a} = {}^{\mathcal{S}}\mathbf{A}_{\mathcal{B}}\,{}^{\mathcal{B}}\mathbf{a}$.
- *Active/Passive Transformation:* Relativity allows transformations to be active (changing object, fixed frame) or passive (changing frame, fixed object).
- *Intrinsic/Extrinsic Rotation:* When discussing rotations around axes, it is necessary to specify the frame to which the rotation axis belongs. Intrinsic rotation uses the axis of $\mathcal{B}$, while extrinsic rotation uses the axis of $\mathcal{S}$.
- *Pre-/Post- Multiplication:* A rotation matrix $\mathbf{A}$ can be applied to a column vector $\mathbf{u}$: cvec n by pre-multiplication ($\mathbf{A}\mathbf{u}$) or to a row vector $\mathbf{u}^{\mathrm{T}}$: rvec n by post-multiplication ($\mathbf{u}^{\mathrm{T}}\mathbf{A}$). To avoid ambiguity, we encapsulate these operations into user-friendly functions using just the vector type (vec n).

3.1   Rotation Matrix

A *rotation matrix* changes the direction of vectors while preserving its magnitude and maintaining the handedness. It often serves as an intermediate form for different $\mathcal{OR}$ methods. These matrices belongs to the special orthogonal group $SO(n, \mathbb{F})$, denoted as $SO(2, \mathbb{R})$ and $SO(3, \mathbb{R})$ in 2D and 3D, respectively.

Orthogonal Matrix: A square matrix $\mathbf{A}$ is termed orthogonal if any of the following conditions hold: (1) Each column consists of unit vectors that are mutually orthogonal. (2) Each row consists of unit vectors that are mutually orthogonal. (3) $\mathbf{A}^\mathsf{T}\mathbf{A} = \mathbf{I}$. We prove the equivalence of these criteria:

```
Definition mcolsOrth {r c} (M : mat r c) : Prop :=
  (∀ i, vunit (mcol M i)) /\ (∀ j k, j <> k -> mcol M j _|_ mcol M k).
Definition morth {n} (M : smat n) : Prop := M\T * M = mat1.
Lemma morth_iff_mcolsOrth : ∀ {n} (M : smat n), morth M <-> mcolsOrth M.
(* row situation is similar *)
```

The key idea to proving the above lemma is to convert matrix orthogonality and multiplication into dot products and show that the dot product equals zero.

Orthogonal matrices possess favorable mathematical properties, including:

1. Orthogonal matrices preserve the dot product.
2. Orthogonal matrices preserve vector lengths.
3. The determinant of an orthogonal matrix is either $+1$ or -1.
4. Orthogonal matrices are always invertible.
5. The inverse of an orthogonal matrix is equal to its transpose.
6. The transpose of an orthogonal matrix is orthogonal.
7. The multiplication of two orthogonal matrices is orthogonal.

These properties are formally stated as the following lemmas.

```
Context {n : nat} (M N : smat n) (a b : vec n).
Lemma morth_keep_dot : morth M -> <M *v a, M *v b> = <a, b>.
Lemma morth_keep_length : morth M -> ||M *v a|| = ||a||. (* ... *)
```

Special Orthogonal Group: The set of n-dimensional orthogonal matrices over $\mathbb{F}$ with determinant 1 constitutes the *special orthogonal group* $SO(n, \mathbb{F})$. Notably, $SO(3, \mathbb{R})$ preserves angles between 3D real vectors and maintains the handedness of a frame (i.e., keeping the positive direction of the third axis) by preserving the cross product. Thereby, using $SO(3, \mathbb{R})$ for orientation transformation guarantees rigid body deformation-free motion.

```
Lemma morth_keep_angle : morth M -> (M *v a) /_ (M *v b) = a /_ b.
Lemma SO3_keep_v3cross : SOnP M -> (M *v a) \x (M *v b) = M *v (a \x b).
```

Here, SOnP indicates a matrix is in $SO(n, \mathbb{F})$, defind as: $\forall \mathbf{M}, \texttt{morth } \mathbf{M} \wedge |\mathbf{M}| = 1$.

3.2   Euler Angles

Euler angles provide an intuitive way to describe rotations, but suffer non-uniqueness due to gimbal lock. There are 24 conventions depending on the choice of rotation axis (intrinsic/extrinsic) and order. We use the notation B for intrinsic rotations about axes in $\mathcal{B}$, S for extrinsic rotations about axes in $\mathcal{S}$, and 1, 2, 3 for axes x, y, z. Thus, S123 represents the extrinsic rotation with the xyz sequence, commonly known as the *roll-pitch-yaw* (RPY) in aerospace, where the angles are usually represented as (ϕ, θ, ψ). Having established the context of Euler angles, we begin with the basic rotation matrix and proceed to discuss its conversion to Euler angles, the singularity issue, and the reverse conversion process.

The *basic rotation matrix* represents rotations around the coordinate axes in a right-hand coordinate system via pre-multiplication and active transformations. The matrices for rotating θ around the x-, y-, or z-axis are:

$$\mathbf{R_x}(\theta) \triangleq \begin{bmatrix} 1 & 0 & 0 \\ 0 & c & -s \\ 0 & s & c \end{bmatrix}, \mathbf{R_y}(\theta) \triangleq \begin{bmatrix} c & 0 & s \\ 0 & 1 & 0 \\ -s & 0 & c \end{bmatrix}, \mathbf{R_z}(\theta) \triangleq \begin{bmatrix} c & -s & 0 \\ s & c & 0 \\ 0 & 0 & 1 \end{bmatrix}. \quad (5)$$

Here, $c = \cos\theta$ and $s = \sin\theta$. The correctness of these matrices can be verified by proving their equivalence to the axis-angle rotation matrices in (10).

Lemma 2 (Euler Angles to Rotation Matrix in S123). *Given rotation angles θ_1, θ_2, and θ_3, the* S123 *rotation matrix is given by:*

$$\mathbf{S123}(\theta_1, \theta_2, \theta_3) \triangleq \begin{bmatrix} c_2c_3 & s_1s_2c_3 - c_1s_3 & c_1s_2c_3 + s_1s_3 \\ c_2s_3 & s_1s_2s_3 + c_1c_3 & c_1s_2s_3 - s_1c_3 \\ -s_2 & s_1c_2 & c_1c_2 \end{bmatrix}, \quad (6)$$

where $c_i = \cos\theta_i$ and $s_i = \sin\theta_i$. In other words, the following equation holds:

$$\mathbf{S123}(\theta_1, \theta_2, \theta_3) = \mathbf{R_z}(\theta_3)\mathbf{R_y}(\theta_2)\mathbf{R_x}(\theta_1). \quad (7)$$

Proof. The algorithm from Eq. (6) is defined as follows:

```
Variable θ1 θ2 θ3 : R. Notation c1 := (cos θ1). (* ... *)
Definition S123 : mat 3 3 := 12m
  [[c2 * c3; s1 * s2 * c3 - c1 * s3; c1 * s2 * c3 + s1 * s3];
   [c2 * s3; s1 * s2 * s3 + c1 * c3; c1 * s2 * s3 - s1 * c3];
   [- s2; s1 * c2; c1 * c2]].
```

The formal description and proof of this lemma are as follows:

```
Theorem S123_spec : S123 = Rz θ3 * Ry θ2 * Rx θ1. Proof. meq; ring. Qed.
```

Lemma 3 (Singularity in S123). *At $\theta = \pm\frac{\pi}{2}$, ϕ and ψ lack unique solutions.*

Proof. We introduce the following lemma to show the singularity's existence:

```
Lemma S123_phi_singular : ∀ φ θ ψ, (θ = π/2 ∨ θ = -π/2) →
  ∀ φ',∃ ψ',S123(φ',θ,ψ') = S123(φ,θ,ψ).
```

The above lemma states that: for any Euler angles (ϕ, θ, ψ), if θ equals $\pm\frac{\pi}{2}$, infinitely many (ϕ', ψ') exist such that the rotation matrix $S123(\phi', \theta, \psi')$ equals $S123(\phi, \theta, \psi)$. The proof relies on trigonometric properties and is omitted.

Lemma 4 (Rotation Matrix to Euler Angles in S123). *For a* S123 *rotation matrix* **T** *with* $\phi, \psi \in (-\pi, \pi)$ *and* $\theta \in (\frac{-\pi}{2}, \frac{\pi}{2})$, *the Euler angles are given by:*

$$
\begin{cases}
\phi(\mathbf{T}) \triangleq \operatorname{atan2}(\mathbf{t}_{32}, \mathbf{t}_{33}) \\
\theta(\mathbf{T}) \triangleq \arcsin(-\mathbf{t}_{31}) \\
\psi(\mathbf{T}) \triangleq \operatorname{atan2}(\mathbf{t}_{21}, \mathbf{t}_{11})
\end{cases}
,
\tag{8}
$$

where $\mathbf{t}_{ij}$ *are elements of* **T**, *and* atan2 *is the four-quadrant inverse tangent which has a range* $(-\pi, \pi]$. *Show that* ϕ, θ, ψ *satisfy* (6).

Proof. We first define the missing atan2 function in CoQ. Then the algorithm (8) is described as follows:

    Definition φ (T : smat 3) := atan2 ((T.3.2)) (T.3.3). (* ... *)

The lemma 4 is formalized as follows:

    Lemma S123_rotMat2euler_spec : **forall** (φ₀ θ₀ ψ₀ : R) (T : smat 3),
    $-\pi < \phi_0 < \pi \rightarrow \frac{-\pi}{2} < \theta_0 < \frac{\pi}{2} \rightarrow -\pi < \psi_0 < \pi \rightarrow$
    T = S123 φ₀ θ₀ ψ₀ → (φ T = φ₀) ∧ (θ T = θ₀) ∧ (ψ T = ψ₀).

Here, T = S123 φ₀ θ₀ ψ₀ indicates that the input matrix **T** is generated from ϕ_0, θ_0, ψ_0 according to S123. The proof is analogous to Lemma 3 and is omitted.

3.3   Axis-Angle

Axis-angle representation describes rotations using a unit vector $\widehat{\mathbf{n}}$ (positive direction of the rotation axis) and angle θ (amount of rotation). This method offers intuitive physical meaning while avoiding gimbal lock. We will derive two key algorithms: vector rotation using axis-angle parameters (Rodrigues' rotation formula) and converting axis-angle parameters to a rotation matrix.

Theorem 1 (Rodrigues' Rotation Formula). *For a unit vector* $\widehat{\mathbf{n}}$ *representing the rotation axis and any angle* θ, *rotating vector* **a** *about* $\widehat{\mathbf{n}}$ *by* θ *yields:*

$$
\mathbf{a}' = \langle \mathbf{a}, \widehat{\mathbf{n}} \rangle \widehat{\mathbf{n}} + \cos\theta (\mathbf{a} - \langle \mathbf{a}, \widehat{\mathbf{n}} \rangle \widehat{\mathbf{n}}) + \sin\theta (\widehat{\mathbf{n}} \times \mathbf{a}).
\tag{9}
$$

Proof. An informal and a formal proof of this theorem are provided below.

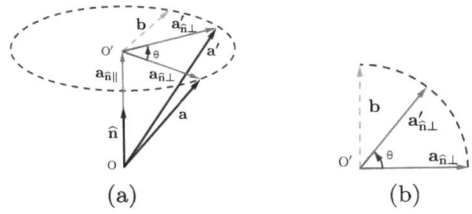

(a)   (b)

Fig. 2. Rotation of vector **a** around axis $\widehat{\mathbf{n}}$ by an angle θ

Informal Proof: Fig. 2 depicts vector $\mathbf{a}$ rotating about axis $\widehat{\mathbf{n}}$ by θ to $\mathbf{a}'$.

Projecting $\mathbf{a}$ onto $\widehat{\mathbf{n}}$, we obtain the parallel component $\mathbf{a}_{\widehat{\mathbf{n}}\|} \triangleq \frac{\langle \mathbf{a}, \widehat{\mathbf{n}} \rangle}{\langle \widehat{\mathbf{n}}, \widehat{\mathbf{n}} \rangle} \widehat{\mathbf{n}} = \langle \mathbf{a}, \widehat{\mathbf{n}} \rangle \widehat{\mathbf{n}}$ and the perpendicular component $\mathbf{a}_{\widehat{\mathbf{n}}\perp} \triangleq \mathbf{a} - \mathbf{a}_{\widehat{\mathbf{n}}\|} = \mathbf{a} - \langle \mathbf{a}, \widehat{\mathbf{n}} \rangle \widehat{\mathbf{n}}$. We then rotate these two components independently to obtain $\mathbf{a}'_{\widehat{\mathbf{n}}\|}$ and $\mathbf{a}'_{\widehat{\mathbf{n}}\perp}$, respectively. Since $\mathbf{a}_{\widehat{\mathbf{n}}\|}$ remains unchanged, (i.e., $\mathbf{a}'_{\widehat{\mathbf{n}}\|} \triangleq \mathbf{a}_{\widehat{\mathbf{n}}\|} = \langle \mathbf{a}, \widehat{\mathbf{n}} \rangle \widehat{\mathbf{n}}$), we focus on $\mathbf{a}'_{\widehat{\mathbf{n}}\perp}$. We define a new vector $\mathbf{b} \triangleq \widehat{\mathbf{n}} \times \mathbf{a}_{\widehat{\mathbf{n}}\perp}$. Note that $\mathbf{b}$ can also be expressed as: $\mathbf{b} = \widehat{\mathbf{n}} \times (\mathbf{a} - \mathbf{a}_{\widehat{\mathbf{n}}\|}) = \widehat{\mathbf{n}} \times \mathbf{a} - \widehat{\mathbf{n}} \times \mathbf{a}_{\widehat{\mathbf{n}}\|} = \widehat{\mathbf{n}} \times \mathbf{a} - \widehat{\mathbf{n}} \times (\langle \mathbf{a}, \widehat{\mathbf{n}} \rangle \widehat{\mathbf{n}}) = \widehat{\mathbf{n}} \times \mathbf{a}$. After rotating $\mathbf{a}_{\widehat{\mathbf{n}}\perp}$ by an angle θ in the plane formed by $\mathbf{a}_{\widehat{\mathbf{n}}\perp}$ and $\mathbf{b}$, we obtain $\mathbf{a}'_{\widehat{\mathbf{n}}\perp}$. According to vector theory, $\mathbf{a}'_{\widehat{\mathbf{n}}\perp} = \cos\theta \mathbf{a}_{\widehat{\mathbf{n}}\perp} + \sin\theta \mathbf{b}$. Thus, we have: $\mathbf{a}'_{\widehat{\mathbf{n}}\perp} = \cos\theta(\mathbf{a} - \langle \mathbf{a}, \widehat{\mathbf{n}} \rangle \widehat{\mathbf{n}}) + \sin\theta(\widehat{\mathbf{n}} \times \mathbf{a})$. Summing the components, $\mathbf{a}' = \mathbf{a}'_{\widehat{\mathbf{n}}\|} + \mathbf{a}'_{\widehat{\mathbf{n}}\perp}$, yields equation (9).

Formal Proof: Equation (9) is formally defined as follows:

```
Definition rotaa (θ : R) (n : vec 3) (a : vec 3) : vec 3 :=
  <a,n> c* n + (cos θ) c* (a - <a,n> c* n) + (sin θ) c* (n×a).
```

The corresponding Theorem 1 is formalized as follows:

```
Theorem rotaa_spec : forall (θ : R) (n a : vec 3),
  let a_proj := vproj a n in  let a_perp := vperp a n in
  let b := n \x a_perp in
  let a_perp' := (cos θ) \.* a_perp + (sin θ) \.* b in
  let a' := a_proj + a_perp' in vunit n -> a' = rotaa θ n a.
Proof. intros; simpl in *.
  assert (a_para = <a,n> c* n)
  assert (a_perp = (a - <a,n> c* n)
  assert (b = n \x a) as H3. (* ... *)
  unfold a'. unfold a_perp'. rewrite H1. rewrite H2. rewrite H3. auto. Qed.
```

Here, some details for subgoals have been omitted for brevity. Note that this formalization closely aligns with the informal derivation presented above and offers a more rigorous and verifiable approach.

Theorem 2 (Axis-Angle to Rotation Matrix). *Given axis-angle parameters $(\widehat{\mathbf{n}}, \theta)$, we define a 3×3 matrix:*

$$\mathbf{R}(\widehat{\mathbf{n}}, \theta) = \mathbf{I}_{3 \times 3} + (\sin\theta)[\widehat{\mathbf{n}}]_{\times} + (1 - \cos\theta)[\widehat{\mathbf{n}}]_{\times}[\widehat{\mathbf{n}}]_{\times}, \tag{10}$$

where $[\mathbf{a}]_{\times}$ denotes skew-symmetric matrix. Then, for any vector $\mathbf{a}$, the rotation of $\mathbf{a}$ by angle θ around $\widehat{\mathbf{n}}$ yields: $\mathbf{a}' = \mathbf{R}(\widehat{\mathbf{n}}, \theta)\mathbf{a}$.

Proof. The CoQ formalization follows. Note that "%R" delimit real numbers.

```
Definition aa2mat (θ n : smat 3) : smat 3 := let N := skew3 n in
  (mat1 + (sin θ) \.* N + (1 - cos θ)
Theorem aa2mat_spec : forall (θ : R) (n a : vec 3),
  vunit n -> rotaa θ n a = (aa2mat θ n) *v a.
```

The proof relying on basic matrix theory, and is omitted due to space limitations.

3.4   Unit Quaternions

This subsection builds on Sect. 2.4 to explore the $\mathcal{OR}$ algorithm using unit quaternions. We present the conversion from axis-angle representation to unit quaternions and rigorously validate the principles of quaternion representation, along with the conversions between unit quaternions and other $\mathcal{OR}$ methods.

Definition 3 (Axis-Angle to Quaternion).

$$\mathrm{aa2q}(\mathbf{n}, \theta) \triangleq \left(\cos \tfrac{\theta}{2}, \sin \tfrac{\theta}{2}\mathbf{n}\right)^{\mathsf{T}} \tag{11}$$

Definition 4 (Rotate Vector Using Unit Quaternions). *Given a unit rotation axis vector $\widehat{\mathbf{n}}$ and a rotation angle θ, the rotation of an arbitrary 3D vector $\mathbf{a}$ around $\widehat{\mathbf{n}}$ by angle θ to obtain the new vector $\mathbf{a}'$ is defined by the formula:*

$$\begin{pmatrix} 0 \\ \mathbf{a}' \end{pmatrix} \triangleq \widehat{\mathbf{q}} \otimes \begin{pmatrix} 0 \\ \mathbf{a} \end{pmatrix} \otimes \widehat{\mathbf{q}}^{-1} = \widehat{\mathbf{q}} \otimes \begin{pmatrix} 0 \\ \mathbf{a} \end{pmatrix} \otimes \overline{\widehat{\mathbf{q}}} \tag{12}$$

where $\widehat{\mathbf{q}} = \mathrm{aa2q}(\widehat{\mathbf{n}}, \theta)$ is a unit quaternion. To specify the rotation operations with quaternions or vectors, we introduce two functions:

$$\mathrm{Q}(\widehat{\mathbf{q}}, \mathbf{p}) \triangleq \widehat{\mathbf{q}} \otimes \mathbf{p} \otimes \overline{\widehat{\mathbf{q}}}, \qquad \mathrm{QV}(\widehat{\mathbf{q}}, \mathbf{a}) \triangleq \mathrm{q2im}\Big(\mathrm{Q}(\widehat{\mathbf{q}}, \mathrm{im2q}(\mathbf{a}))\Big). \tag{13}$$

where Q takes and outputs quaternions, and QV takes and outputs 3D real vectors. They are formalized as qrot and qrotv, respectively.

   Definition qrot (q p : quat) : quat := q * p * qconj q.

   Definition qrotv (q : quat) (a:vec 3) : vec 3 := q2im (qrot q (im2q a)).

The function Q and QV have the following properties:

(a) Q preserves the scalar component: $\mathrm{W}(\mathrm{Q}(\widehat{\mathbf{q}}, \mathbf{p})) = \mathrm{W}(\mathbf{p})$.
(b) QV preserves the dot product: $\langle \mathrm{QV}(\widehat{\mathbf{q}}, \mathbf{a}), \mathrm{QV}(\widehat{\mathbf{q}}, \mathbf{b}) \rangle = \langle \mathbf{a}, \mathbf{b} \rangle$.
(c) QV preserves the length: $\|\mathrm{QV}(\widehat{\mathbf{q}}, \mathbf{a})\| = \|\mathbf{a}\|$.
(d) QV preserves the addition: $\mathrm{QV}(\widehat{\mathbf{q}}, \mathbf{a} + \mathbf{b}) = \mathrm{QV}(\widehat{\mathbf{q}}, \mathbf{a}) + \mathrm{QV}(\widehat{\mathbf{q}}, \mathbf{b})$.
(e) QV preserves the scalar multiplication: $\mathrm{QV}(\widehat{\mathbf{q}}, x\mathbf{a}) = x\mathrm{QV}(\widehat{\mathbf{q}}, \mathbf{a})$.

Here is some of the description of these properties:

   Variable q : quat. **Hypothesis** Hq : qunit q.

   Lemma qrot_keep_w: **forall** p:quat, W (qrot q p) = W p.

   Lemma qrot_keep_dot: **forall** a b:vec 3, <qrotv q a, qrotv q b> = <a, b>.

Lemma 5 (Decomposition of Quaternion Multiplication). *For any two unit vectors $\mathbf{a}$ and $\mathbf{b}$, their multiplication satisfies the following equation:*

$$\begin{pmatrix} \langle \mathbf{a}, \mathbf{b} \rangle \\ \mathbf{a} \times \mathbf{b} \end{pmatrix} = \begin{pmatrix} 0 \\ \mathbf{b} \end{pmatrix} \otimes \overline{\begin{pmatrix} 0 \\ \mathbf{a} \end{pmatrix}}. \tag{14}$$

Proof. The relations in (14) can be formally stated as follows:

Definition ab2q (a b : vec 3) : quat := si2q (<a, b>) (a \x b).

Definition ab2q' (a b : vec 3) : quat := (im2q b) * (qconj (im2q a)).

Lemma ab2q_eq_ab2q' : **forall** a b, ab2q a b = ab2q' a b.

Proof. intros. lqa. **Qed.**

Here, ab2q is the geometric interpretation, while ab2q' is the algebraic one.

Lemma 6 (Invariant of Rotation by Quaternion). *Given two unit vectors* **a** *and* **b**, *let* $\widehat{\mathbf{q}} = \begin{pmatrix} \langle \mathbf{a}, \mathbf{b} \rangle \\ \mathbf{a} \times \mathbf{b} \end{pmatrix}$, *and let* **c** *be the result of* QV($\widehat{\mathbf{q}}$, **a**), *show* $\begin{pmatrix} \langle \mathbf{b}, \mathbf{c} \rangle \\ \mathbf{b} \times \mathbf{c} \end{pmatrix} = \widehat{\mathbf{q}}$.

Proof. The lemma is formally stated as:

Lemma ab2q_eq : **forall** (a b : vec 3), **let** q := ab2q a b **in**

let c := qrotv q a **in** vunit a -> vunit b -> ab2q b c = q.

The proof of this lemma relies on properties of quaternion multiplication and conjugation, which are omitted here due to space limitations. Note that this lemma will be used multiple times in the Theorem 3.

Theorem 3 (Quaternion Rotation Algorithm). *Given an axis-angle parameter* $(\widehat{\mathbf{n}}, \theta)$, *where* $\theta \in (0, 2\pi)$, *and a unit quaternion* $\widehat{\mathbf{q}} \triangleq (\cos \frac{\theta}{2}, \sin \frac{\theta}{2} \widehat{\mathbf{n}})^{\mathsf{T}}$, *applying the transformation defined by Eq. (13) with* $\widehat{\mathbf{q}}$ *to any 3D vector* **v** *results in* **v**′. *We demonstrate that this transformation rotates* **v** *about axis* $\widehat{\mathbf{n}}$ *by angle* θ *to yield* **v**′. *As shown in Fig. 3(a), we prove two properties: (1) Magnitude Preservation:* $\|\mathbf{v}'\| = \|\mathbf{v}\|$; *(2) Angle Preservation: The angle between the perpendicular component of* **v** *and* **v**′ *onto* $\widehat{\mathbf{n}}$ *is* θ, *that is* $\mathbf{v}_{\widehat{n}\perp} \angle \mathbf{v}'_{\widehat{n}\perp} = \theta$.

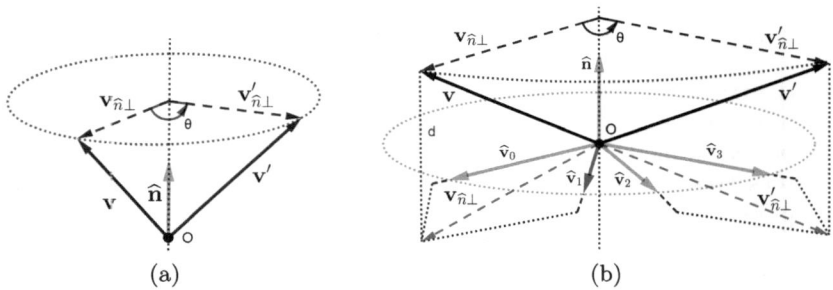

(a)   (b)

Fig. 3. Rotate **v** to **v**′ about $\widehat{\mathbf{n}}$ by θ degree

Proof. The Magnitude Preservation is ensured by the property c of QV. Formally:

Let v' : vec 3 := qrotv q v.

Fact vlen_vv' : ||v'|| = ||v||.

Proof. unfold v',v. **rewrite** qrot_keep_vlen; **auto. Qed.**

Next, we analyze the Angle Preservation. Since the magnitude and direction of $\mathbf{v}$ are arbitrary, it is not convenient for analysis. We decompose it into a linear combination represented by a set of basis vectors and discuss the transformation of each component. As shown in Fig. 3(b), we project $\mathbf{v}$ and $\mathbf{v}'$ onto a plane d perpendicular to $\widehat{\mathbf{n}}$.

First, we construct basis vectors to represent $\mathbf{v}$. Assume that there exists a pair of linearly independent unit vectors $\widehat{\mathbf{v}}_0$ and $\widehat{\mathbf{v}}_1$ in the plane d perpendicular to $\widehat{\mathbf{n}}$ such that $\widehat{\mathbf{v}}_0 \angle \widehat{\mathbf{v}}_1 = \frac{\theta}{2}$. This assumption is reasonable because $0 < \frac{\theta}{2} < \pi$, so $\widehat{\mathbf{v}}_0$ and $\widehat{\mathbf{v}}_1$ must be linearly independent. Thus, $\widehat{\mathbf{v}}_0$, $\widehat{\mathbf{v}}_1$, and $\widehat{\mathbf{n}}$ form basis vectors in 3D space, and $\mathbf{v}$ can be decomposed as $\mathbf{v} = s_0\widehat{\mathbf{v}}_0 + s_1\widehat{\mathbf{v}}_1 + s_2\widehat{\mathbf{n}}$ for some s_0, s_1, s_2. Formally, these assumptions are stated as:

Variables $(\theta : R)$ (n v0 v1 : vec 3) (s0 s1 s2 : R).

Hypotheses (Hθ: $0 < \theta < 2 *$ PI) (Hv0: vunit v0) (Hv1: vunit v1) (H01n: vnorm (v0×v1) = n) (H01: v0 $\angle$ v1 = θ/2) (Hs0: s0<>0) (Hs1: s1<>0).

Let v : vec 3 := (s0 \.* v0 + s1 \.* v1 + s2 \.* n)

Let v' : vec 3 := qrotv q v.

We can prove the following facts:

$$\begin{pmatrix} \langle \widehat{\mathbf{v}}_0, \widehat{\mathbf{v}}_1 \rangle \\ \widehat{\mathbf{v}}_0 \times \widehat{\mathbf{v}}_1 \end{pmatrix} = \widehat{\mathbf{q}}, \quad \widehat{\mathbf{v}}_0 \perp \widehat{\mathbf{n}}, \quad \widehat{\mathbf{v}}_1 \perp \widehat{\mathbf{n}}. \tag{15}$$

Next, we use $\widehat{\mathbf{q}}$ to transform $\widehat{\mathbf{v}}_0$ to $\widehat{\mathbf{v}}_2$, and obtain:

$$\begin{pmatrix} \langle \widehat{\mathbf{v}}_1, \widehat{\mathbf{v}}_2 \rangle \\ \widehat{\mathbf{v}}_1 \times \widehat{\mathbf{v}}_2 \end{pmatrix} = \widehat{\mathbf{q}}, \quad \widehat{\mathbf{v}}_2 \perp \mathbf{v}, \quad \widehat{\mathbf{v}}_1 \angle \widehat{\mathbf{v}}_2 = \theta/2, \quad \widehat{\mathbf{v}}_0 \angle \widehat{\mathbf{v}}_2 = \theta. \tag{16}$$

Then, we use $\widehat{\mathbf{q}}$ to transform $\widehat{\mathbf{v}}_1$ to $\widehat{\mathbf{v}}_3$, and obtain:

$$\begin{pmatrix} \langle \widehat{\mathbf{v}}_2, \widehat{\mathbf{v}}_3 \rangle \\ \widehat{\mathbf{v}}_2 \times \widehat{\mathbf{v}}_3 \end{pmatrix} = \widehat{\mathbf{q}}, \quad \widehat{\mathbf{v}}_3 \perp \mathbf{v}, \quad \widehat{\mathbf{v}}_2 \angle \widehat{\mathbf{v}}_3 = \theta/2, \quad \widehat{\mathbf{v}}_1 \angle \widehat{\mathbf{v}}_3 = \theta. \tag{17}$$

Then, using $\widehat{\mathbf{q}}$ to transform $\widehat{\mathbf{n}}$ leaves it unchanged: $\mathrm{QV}(\widehat{\mathbf{q}}, \widehat{\mathbf{n}}) = \widehat{\mathbf{n}}$. Finally, we can begin to prove that the angle between $\mathbf{v}_{\widehat{\mathbf{n}}\perp}$ and $\mathbf{v}'_{\widehat{\mathbf{n}}\perp}$ is θ. Formally:

Fact vangle_vv' : vperp v n $\angle$ vperp v' n = θ.

Once unfold the definitions of v and v', the goal becomes to:

(vperp (s0 \.* v0 + s1 \.* v1 + s2 \.* n) n) $\angle$
   (vperp (qrotv q (s0 \.* v0 + s1 \.* v1 + s2 \.* n)) n) = θ

Using the linearity property of QV and projection, the goal simplifies to:

(s0 \.* vperp v0 n + s1 \.* vperp v1 n + s2 \.* vperp n n) $\angle$
   (s0 \.* vperp v2 n + s1 \.* vperp v3 n + s2 \.* vperp n n) = θ

By using the properties of projection, the goal becomes to:

(s0 \.* v0 + s1 \.* v1) $\angle$ (s0 \.* v2 + s1 \.* v3) = θ

Since linear operations preserve angles, our goals are:

v0 $\angle$ v2 = θ v0 $\angle$ v1 = v2 $\angle$ v3

Both goals are readily established from Eqs. (16) and (17), thus concludes the proof (omitting intermediate steps for brevity).

Next, we provide a complete theorem statement:

Theorem qrot_spec : **forall** (v0 v1 n : vec 3) (θ : R) (s0 s1 s2 : R)

  let q : quat := aa2q (θ, n) **in**

  let v : vec 3 := (s0 \.* v0 + s1 \.* v1 + s2 \.* n)

  let v' : vec 3 := qrotv q v **in**

  vunit v0 -> vunit v1 -> 0 < θ < 2*PI ->

  vnorm (v0 × v1) = n -> v0∠v1 = θ/2 -> s0<>0 -> s1<>0 ->

  (||v'|| = ||v||) /\ vperp v n ∠ vperp v' n = θ.

The proof of this theorem is as follows:

Proof. split; [**apply** vlen_vv'| **apply** vangle_vv']; **auto**. **Qed**.

Here, we immediately finish the proof with the established facts above.

Additionally, we used an alternative method to prove this theorem. We demonstrated the equivalence between the results from the quaternion rotation algorithm (13) and the axis-angle formula (9). Formally:

Lemma qrot_spec_byAxisAngle : **forall** (θ : R) (n v : vec 3),

  vunit n -> qrotv (aa2q (θ,n)) v = rotaa θ n v.

The proof is essentially computational and is omitted here.

Lemma 7 (Multiple Rotations by Unit Quaternions). *Given axis-angle parameters* $(\widehat{\mathbf{n}}_1, \theta_1)$ *and* $(\widehat{\mathbf{n}}_2, \theta_2)$ *represented by unit quaternions* $\widehat{\mathbf{q}}_1$ *and* $\widehat{\mathbf{q}}_2$, *respectively, for any vector* $\mathbf{v}$, *the result of rotating* $\mathbf{v}$ *with* $\widehat{\mathbf{q}}_1$ *followed by* $\widehat{\mathbf{q}}_2$ *is:*

$$\begin{pmatrix} 0 \\ \mathbf{v}'' \end{pmatrix} = \widehat{\mathbf{q}} \otimes \begin{pmatrix} 0 \\ \mathbf{v} \end{pmatrix} \otimes \overline{\widehat{\mathbf{q}}}, \quad \text{where } \widehat{\mathbf{q}} = \widehat{\mathbf{q}}_2 \otimes \widehat{\mathbf{q}}_1. \tag{18}$$

Proof. The derivation proceeds as follows:

$$\begin{pmatrix} 0 \\ \mathbf{v}'' \end{pmatrix} \triangleq \widehat{\mathbf{q}}_2 \otimes \begin{pmatrix} 0 \\ \mathbf{v}' \end{pmatrix} \otimes \overline{\widehat{\mathbf{q}}_2} = \widehat{\mathbf{q}}_2 \otimes \left(\widehat{\mathbf{q}}_1 \otimes \begin{pmatrix} 0 \\ \mathbf{v} \end{pmatrix} \otimes \overline{\widehat{\mathbf{q}}_1} \right) \otimes \overline{\widehat{\mathbf{q}}_2}$$

$$= (\widehat{\mathbf{q}}_2 \otimes \widehat{\mathbf{q}}_1) \otimes \begin{pmatrix} 0 \\ \mathbf{v} \end{pmatrix} \otimes (\overline{\widehat{\mathbf{q}}_1} \otimes \overline{\widehat{\mathbf{q}}_2}) = (\widehat{\mathbf{q}}_2 \otimes \widehat{\mathbf{q}}_1) \otimes \begin{pmatrix} 0 \\ \mathbf{v} \end{pmatrix} \otimes (\overline{\widehat{\mathbf{q}}_2 \otimes \widehat{\mathbf{q}}_1})$$

The formal proof is provided below:

Lemma qrot_twice : ∀ (q1 q2:quat) (v:vec 3), q1 ≠ qzero -> q2 ≠ qzero ->
qrot q2 (qrot q1 (im2q v)) = qrot (q2 * q1) (im2q v).

Proof. unfold qrot. **rewrite** qinv_qmul, !qmul_assoc; **auto**. **Qed**.

Lemma 8 (Unit Quaternions to Rotation Matrix). *Given a unit quaternion* $\widehat{\mathbf{q}} = (w, x, y, z)^{\mathrm{T}}$ *representing the rotation from* $\mathcal{B}$ *to* $\mathcal{S}$, *for any 3D vector* $\mathbf{a}$, *we have* ${}^{\mathcal{S}}\mathbf{a} = \mathrm{QV}(\widehat{\mathbf{q}}, {}^{\mathcal{B}}\mathbf{a})$. *Show that the matrix* $\mathbf{R}(\widehat{\mathbf{q}})$:

$$\mathbf{R}(\widehat{\mathbf{q}}) \triangleq \begin{bmatrix} w^2 + x^2 - y^2 - z^2 & 2(xy - wz) & 2(xz + wy) \\ 2(xy + wz) & w^2 - x^2 + y^2 - z^2 & 2(yz - wx) \\ 2(xz - wy) & 2(yz + wx) & w^2 - x^2 - y^2 + z^2 \end{bmatrix}, \quad (19)$$

satisfy $^{\mathcal{S}}\mathbf{a} = \mathbf{R}(\widehat{\mathbf{q}})^{\mathcal{B}}\mathbf{a}$.

Proof. From the equation $^{\mathcal{S}}\mathbf{a} = \mathtt{QV}(\widehat{\mathbf{q}}, {}^{\mathcal{B}}\mathbf{a})$, we know that $(0, {}^{\mathcal{S}}\mathbf{a})^{\mathsf{T}} = \mathtt{Q}(\widehat{\mathbf{q}}, (0, {}^{\mathcal{B}}\mathbf{a})^{\mathsf{T}})$. Expanding this definition, we have:

$$(0, {}^{\mathcal{S}}\mathbf{a})^{\mathsf{T}} = \widehat{\mathbf{q}} \otimes (0, {}^{\mathcal{B}}\mathbf{a})^{\mathsf{T}} \otimes \overline{\widehat{\mathbf{q}}}$$

According to Eqs. (3), as well as the associativity of quaternion and matrix multiplications, we have:

$$(0, {}^{\mathcal{S}}\mathbf{a})^{\mathsf{T}} = \mathtt{ML}(\widehat{\mathbf{q}})\mathtt{MR}(\overline{\widehat{\mathbf{q}}})(0, {}^{\mathcal{B}}\mathbf{a})^{\mathsf{T}}$$

Here, $\mathtt{ML}(\widehat{\mathbf{q}})\mathtt{MR}(\overline{\widehat{\mathbf{q}}})$ yields a 4×4 matrix, where the right bottom part should be exactly equal to $\mathbf{R}(\widehat{\mathbf{q}})$. The formalization is below:

```
Lemma q2m_spec q v : qunit q -> qrotv q v = (q2m q) *v v.
Proof. intros. hnf. destruct q. lma; ra. Qed.
```

Here, q2m is the function of (19), and the proof is mainly automated.

Lemma 9 (Rotation Matrix to Unit Quaternions). *Given an orthogonal matrix $\mathbf{A}_{3\times3}$ representing the rotation from $\mathcal{B}$ to $\mathcal{S}$ such that for any 3D vector $\mathbf{a}$, $^{\mathcal{S}}\mathbf{a} = \mathbf{A}^{\mathcal{B}}\mathbf{a}$. Show that the quaternion $\widehat{\mathbf{q}}$, defined by its components as:*

$$\widehat{\mathbf{q}} \triangleq \begin{pmatrix} s_0 \frac{1}{2}\sqrt{1 + a_{11} + a_{22} + a_{33}} \\ s_1 \frac{1}{2}\sqrt{1 + a_{11} - a_{22} - a_{33}} \\ s_2 \frac{1}{2}\sqrt{1 - a_{11} + a_{22} - a_{33}} \\ s_3 \frac{1}{2}\sqrt{1 - a_{11} - a_{22} + a_{33}} \end{pmatrix}, \quad (20)$$

satisfies $^{\mathcal{S}}\mathbf{a} = \mathtt{QV}(\widehat{\mathbf{q}}, {}^{\mathcal{B}}\mathbf{a})$. Here, the signs s_0, s_1, s_2, s_3 are determined by: $s_0 \triangleq \pm 1, s_1 \triangleq s_0 \mathtt{Sgn}(a_{32} - a_{23}), s_2 \triangleq s_0 \mathtt{Sgn}(a_{13} - a_{31}), s_3 \triangleq s_0 \mathtt{Sgn}(a_{21} - a_{12})$. Thus, there are two potential quaternions representing the rotation: $\widehat{\mathbf{q}}$ and $-\widehat{\mathbf{q}}$.

The algorithm's correctness is based on the properties of orthogonal matrices, details of which are omitted due to space limitations.

4   Evaluation

To facilitate user access to verified algorithms, we defined direct interfaces in the OrienRepr.v module, supporting conversions between various OR parameters, as shown in Fig. 4. Additionally, we extracted these algorithms to OCaml and conducted a simple experiment. Given the Euler angles (0.1,1.2,0.8),

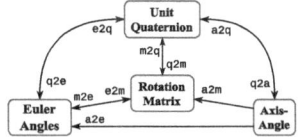

Fig. 4. User functions for conversions between different ORs.

e2m generated the rotation matrix $\begin{bmatrix} 0.252 & -0.648 & 0.717 \\ 0.259 & 0.759 & 0.595 \\ -0.932 & 0.036 & 0.360 \end{bmatrix}$. This was converted to the quaternion $(0.770; -0.181; 0.535; 0.295)$ using m2q. Next, q2a converted this to the Axis-angle $(-0.284; 0.839; 0.462; 1.383)$, and a2e returned to Euler angles $(0.099, 1.200, 0.8)$. These results match those from an online tool [8].

5   Related Works

The fields of $\mathcal{OR}$ and $\mathcal{PR}$ algorithms have a rich history of research, with ongoing contributions from both theoretical analysis and formalization.

Theoretical analysis of $\mathcal{OR}$ and $\mathcal{PR}$ algorithms has seen significant contributions from textbooks such as Fletcher Dunn and Ian Parberry's comprehensive guide to 3D graphics and game development [6], Peter Corke's influential work in robotics and computer vision [4], Quan Quan's exploration of multicopter orientation issues [12], and James Diebel's extensive models of rigid body orientation [5]. These sources, though thorough, generally lack formal verification.

In the realm of formal verification, several works have inspired our research approach. Pham formalizes concepts related to similar triangles and common theorems in elementary plane geometry using using CoQ in [11]. Wu Aixuan et al. formalize screw theory and some kinematic problems in general 6R robotic arms using HOL4 in [16]. Zhenwei Ma et al. verify equations of composite transformation matrices between coordinate systems in [9]. Guojun Xie et al. establish forward kinematic models of robots under the DH convention based on CoQ, modeling common link structures in industrial robots in [17,18]. Andrea Gabrielli and Marco Maggesi formalize quaternions using HOL Light and demonstrate some significant applications in [7]. However, these research works focus on individual models and lack a comprehensive analysis of all $\mathcal{OR}$ models.

It is worth mentioning that Reynald Affeldt and Cyril Cohen have formalized rigid body transformations using the MathComp library, including basic 3D geometry theories such as lines and frames, exponential and skew-symmetric matrices, quaternions, homogeneous representations [1]. Although their work is outstanding, ours has some distinctive features. Firstly, we derive the basic formulas of axis-angle rotation and quaternion rotation from a geometry perspective, as seen in Theorems 1, 2 and 3, rather than using purely algebraic methods. Secondly, compared to the MathComp library, our self-built vector and matrix libraries have lightweight construction, unified naming, and concise notation (while MathComp employs complex notations, SSR syntax, and involves 14 layers to construct the matrix type, increasing comprehension difficulty). Thirdly, our vector operations are very intuitive, where dot product and cross product are direct element-wise operations, while in Affeldt and Cohen's definition, the dot product involves matrix multiplication followed by element extraction, and the cross product involves scalar multiplication of basis vectors using determinants. The fundamental reason is that we provide a vector library, whereas MathComp's vectors are matrices of one row or one column and do not have true vector types. Additionally, our matrix library allows for direct expression evaluation to elements in CoQ, whereas MathComp's library yields expressions that cannot be readily streamlined, impeding users' intuitive comprehension of results.

6   Conclusions

Control systems are designed to achieve desired poses or trajectories for objects or systems, involving both translation and rotation. The representation of rotation ($\mathcal{OR}$) requires careful consideration due to its complexity. Several mathematical models exist for $\mathcal{OR}$, each with its own mathematical underpinnings. Mathematicians and engineers use these models to construct algorithms for practical applications. However, concerns remain regarding the functionality and potential design flaws within these algorithms.

In this paper, we leveraged the CoQ PROOF ASSISTANT to formally describe and verify the mathematical models and algorithms for $\mathcal{OR}$. This rigorous verification ensures the correctness of the algorithms by systematically verifing each step of the underlying mathematical reasoning. Furthermore, by explicitly presenting the connections between various definitions and properties, we enhance the accessibility of these algorithms for software developers unfamiliar with the relevant engineering domain. We also performed code extraction from the verified algorithms, obtaining satisfactory results, which are not detailed here due to space limitations. Our work is available in the open-source project **OrienRepr** (https://zhengpushi.github.io/projects/OrienRepr).

Our main contributions are as follows: (1) Establishing a formal software library for fundamental mathematical tools like vectors, matrices, and quaternions, with a hierarchical structure for scalability and reusability. (2) Formalizing algorithms and specifications for widely used $\mathcal{OR}$ methods, including rotation matrices, Euler angles, axis-angle, and unit quaternions. (3) Conducting complete derivations and verifications of critical algorithms (such as Theorems 1, 2 and 3), demonstrating the use of our library in CoQ for modeling and verifying geometric and kinematic problems.

Further research opportunities arising from our work include: (1) Formalizing additional kinematic algorithms, such as pose representation that integrates translation and rotation motions. (2) Formalizing further matrix theories, including eigenvalues, matrix decomposition, and generalized inverses. (3) Translating our verified algorithms into user-friendly software libraries for programmers.

Continuing this research can contribute to the advancement of reliability and efficiency in control systems, impacting the broader fields of mathematics, engineering, and software development.

Acknowledgment. We extend our thanks to seminar colleagues for their insightful discussions and reviewers for their constructive feedback, which enhanced our work. Our previous work on the FinMatrix [13] project greatly contributed to this paper.

References

1. Affeldt, R., Cohen, C.: Formal foundations of 3D geometry to model robot manipulators. In: Proceedings of the 6th ACM SIGPLAN Conference on Certified Programs and Proofs, pp. 30–42. ACM, Paris (2017). https://doi.org/10.1145/3018610.3018629

2. Chen, G.: Formalized mathematics and proof engineering (in chinese) **12**(9) (2016). https://dl.ccf.org.cn/article/articleDetail.html?id=3738875402700800

3. Coq Development Team: Coq Proof Assistant. https://coq.inria.fr

4. Corke, P.: Robotics. Vision and Control. Springer, Heidelberg (2017). https://doi.org/10.1007/978-3-319-54413-7

5. Diebel, J.: Representing attitude: Euler angles, unit quaternions, and rotation vectors (2006). https://api.semanticscholar.org/CorpusID:16450526

6. Dunn, F., Parberry, I.: 3D Math Primer for Graphics and Game Development. CRC Press (2011). https://gamemath.com/

7. Gabrielli, A., Maggesi, M.: Formalizing basic quaternionic analysis. In: Ayala-Rincón, M., Muñoz, C.A. (eds.) Interactive Theorem Proving, p. 16. Springer, Heidelberg (2017). https://doi.org/10.1007/978-3-319-66107-0_15

8. Gaschler, A.: 3d rotation converter (2024). https://www.andre-gaschler.com/rotationconverter/

9. Ma, Z., Chen, G.: Matrix formalization based on coq record. Comput. Sci. **46**(7), 139 (2019). https://doi.org/10.11896/j.issn.1002-137X.2019.07.022

10. Mahboubi, A., Tassi, E.: Mathematical Components. Zenodo (2022). https://doi.org/10.5281/zenodo.7118596

11. Pham, T.M.: Similar triangles and orientation in plane elementary geometry for coq-based proofs. In: Proceedings of the 2010 ACM Symposium on Applied Computing, pp. 1268–1269. ACM (2010). https://doi.org/10.1145/1774088.1774358

12. Quan, Q.: Introduction to Multicopter Design and Control. Springer, Heidelberg (2017). https://doi.org/10.1007/978-981-10-3382-7

13. Shi, Z.: Finmatrix project (2024). https://zhengpushi.github.io/projects/FinMatrix

14. Shi, Z., Chen, G.: Integration of multiple formal matrix models in coq. In: SETTA 2022, Beijing, China, 27–29 October 2022, vol. 13649, pp. 169–186. Springer, Cham (2022). https://doi.org/10.1007/978-3-031-21213-0_11

15. Shi, Z., Xie, G., Chen, G.: Coqmatrix: formal matrix library with multiple models in coq. J. Syst. Architect. **143**, 102986 (2023). https://doi.org/10.1016/j.sysarc.2023.102986

16. Wu, A., et al.: Formal kinematic analysis of a general 6r manipulator using the screw theory. Math. Probl. Eng. **2015**, e549797 (2015). https://doi.org/10.1155/2015/549797

17. Xie, G., Yang, H., Deng, H., Shi, Z., Chen, G.: Formal verification of robot rotary kinematics. Electronics **12**(2), 36 (2023). https://doi.org/10.3390/electronics12020369

18. Xie, G., Yang, H., Shi, Z., Chen, G.: Formal verification of robot forward kinematics based on dh coordinate system. J. Softw. **35**(9) (2024). https://doi.org/10.13328/j.cnki.jos.007131

Reactive Graphs in Action

David Tinoco[1], Alexandre Madeira[1], Manuel A. Martins[1],
and José Proença[2]([✉])

[1] CIDMA, Department of Mathematics, University of Aveiro, Aveiro, Portugal
[2] CISTER, Faculty of Sciences, University of Porto, Porto, Portugal
jose.proenca@fc.up.pt

Abstract. *Reactive graphs* are transition structures whereas edges become active and inactive during its evolution, that were introduced by Dov Gabbay from a mathematical's perspective. This paper presents Marge (https://fm-dcc.github.io/MARGe), a web-based tool to visualise and analyse reactive graphs enriched with labels. Marge animates the operational semantics of reactive graphs and offers different graphical views to provide insights over concrete systems. We motivate the applicability of reactive graphs for adaptive systems and for featured transition systems, using Marge to tighten the gap between the existing theoretical models and their usage to analyse concrete systems.

1 Introduction and Motivation

A reactive graph is a transition structure that updates its transitions along its execution. This concept has been introduced by Dov Gabbay in [12]. It generalizes the static notion of a graph by incorporating high-order edges that capture updates on the accessibility relations. The notion of reactivity for these structures is not coined only in the standard sense of Harel and Pnueli [16], as systems that react to their environment and are not meant to terminate, but as systems whose accessibility relation is a result of the transformations induced by the transitions executed so far.

Let us consider the model of a simple vending machine in Fig. 1 to motivate reactive graphs. Edges in reactive graphs can be active or inactive, and only transitions involving active edges can be executed. All edges in our vending machine are active, except one with a dotted line. Executing a transition triggers an update on the set of active and inactive edges.

The machine *VM* in Fig. 1 can receive from the user at most 1€. The arrows between states represent *ground edges*, which are labelled with actions; the others

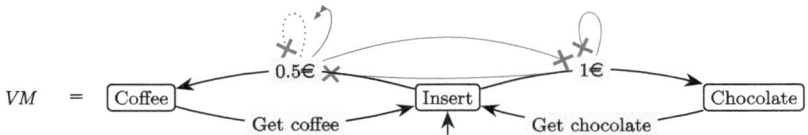

Fig. 1. A reactive graph of a vending machine offering two different products

arrows represent *hyper edges*, i.e., edges that can activate (➤➤) and deactivate
(─×) edges. When the action 1€ is performed some edges are deactivated and the
machine goes to the *Chocolate* state. More specifically, both edges labelled by
0.5€ and 1€ are deactivated. Executing instead the edge labelled by 0.5€ would
enable a deactivating edge, and executing it a second time would deactivate it.

A solid supporting theory for these models, including proposals of specification logics, has been studied in the last years (e.g. [14,15]) and was summed
up in the book [13]. The main advantage of reactive graphs with respect to
traditional labelled transition systems (LTS) is the compact representation of
dynamic systems. Gabbay showed the encoded LTS of a given reactive system
can have an exponentially larger number of states [13, Prop. 8.8]. For example,
our vending machine can be expressed with an LTS with seven states, depicted in
Fig. 2. In a larger example borrowed from Cordy et al. [8, Fig. 1], described in a
companion report [21], the encoded LTS has 7x more states and 4.6x more edges
(including hyper edges). Furthermore, we believe that many examples become
easier to understand and to maintain with reactive graphs than with traditional
LTS.

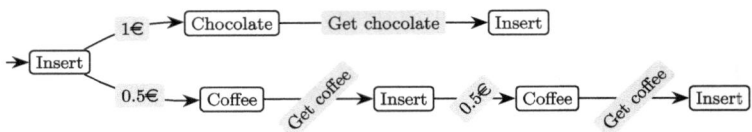

Fig. 2. The LTS of the vending machine in Fig. 1

Reconfigurable Systems and Variability. Not all scenarios and application
domains can exploit the advantages of modelling by reactive graphs. The benefits of compactness and being more intuitive are more evident when analysing
reconfigurable systems. These are systems that operate in different modes
of execution, e.g., an operating systems that supports users with different per-
missions or different performance modes (e.g. an ocean exploring robot that
adjusts its behaviour based on the distance of its base, its energy levels, and its
environmental conditions).

Work on *software product lines* (SPL) focuses mainly on **configurable systems**, i.e., how to develop, maintain, and reason over families of software that
share many commonalities. Feature transition systems [3,8] are structures often
used to model such systems, which are enriched with annotations over features,
allowing an initial configuration of the variant to select which transitions are
active. A branch of SPL focuses on *dynamic SPL* [7] addressing *reconfigurable
and self-adaptive systems*, in which the configuration can change over time, which
relates to our reactive graphs.

This paper introduces Marge: an open-source web-based prototype tool
designed for the visualisation and analysis of labelled reactive graphs. It includes
several examples, both from the literature on reactive graphs [13] and from work

on dynamic SPL [8]. The goal of Marge is to help increasing the adoption of reactive graphs, providing insights over the capabilities and challenges of modelling reconfigurable systems with reactive graphs, and to expose these to different domains to increase it applicability. Currently Marge does not aim at contributing directly to the community on adaptative SPLs, since it still misses a more user-friendly specification language and mechanisms to support larger systems. Marge provides support to: (1) *visualise* reactive graphs, (2) *animate* its operational semantics, (3) *explore* the full state-space of its underlying LTS, (4) *verify* properties such as deadlocks and conflicts, and (5) *compare* reactive graphs using an observational equivalence.

2   Multi-actions Reactive Graphs

A *multi-action reactive graph*, or simply *reactive graph*, is a labelled transition system with transitions enriched with a reaction that activates and deactivates transitions, defined formally below.

Definition 1 (Reactive Graph). *A* Multi-Action Reactive Graph *is a tuple* $M = (W, Act, E, \to, \twoheadrightarrow, \rightarrowtail, \overline{\cdot}, w_0, \alpha_0)$ *where:*

- $W \neq \varnothing$ *is the* set *of states;* Act *is the* set *of actions;* E *is the* set *of edges;*
- $\to \ \subseteq \ W \times Act \times W$ *is the set of ground edges;* $\twoheadrightarrow \ \subseteq \ E \times E$ *is the set of activating edges;* $\rightarrowtail \ \subseteq \ E \times E$ *is the set of deactivating edges;* $\overline{\cdot} : E \longrightarrow (\to \cup \twoheadrightarrow \cup \rightarrowtail)$ *is an injective function that maps edges in E to their internal details;*
- $w_0 \in W$ *is the* initial state; $\alpha_0 \in E$ *is the set of* initially active edges.

Notation. Recall the vending machine in Fig. 1. When formalising this as a reactive graph we say that: (i) *Coffee* belongs to W (among others), (ii) *0.5€* is an action in Act, $\langle Insert, 0.5\text{€}\rangle$, *Coffee* is a ground edge in $\to$, (iii) $\langle e^\dagger, e^\dagger\rangle$ is a deactivating edge in $\rightarrowtail$ where $\overline{e^\dagger} = \langle Insert, 0.5\text{€}, Coffee\rangle$, (iv) $w_0 = Insert$, and (v) α_0 is the set of edges in E without the deactivating edge $\langle e^\dagger, e^\dagger\rangle$.

A reactive graph has an initial state w_0 and an initial set of active edges α_0. Evolving a reactive graph means transitioning to a new state, connected by an enabled ground edge from w_0, and updating the set of active edges. We start by formalising the set of activate and deactivate edges by another given edge, and then formalise the evolution of reactive graphs.

Definition 2 (Activation and Deactivation). *Given a reactive graph M, an edge $e \in E_M$ and a set of active edges $\alpha \subseteq E_M$, we define the set of edges activated by e (resp. deactivated by e), written* $\mathsf{on}(e, \alpha)$ *(resp.* $\mathsf{off}(e, \alpha)$*) as follows.*

$$\mathsf{from}(e_s) = \{e \mid \exists e_t \cdot \overline{e} = (e_s, e_t)\}$$

$$\mathsf{from}^*(e, \alpha) = \bigcup\nolimits_{r \in (\mathsf{from}(e) \cap \alpha)} \mathsf{from}^*(r, \alpha \backslash \{e\}) \cup \{r\}$$

$$\mathsf{on}(e, \alpha) = \{e_t \mid e_{trg} \in \mathsf{from}^*(e, \alpha) \wedge \exists e_s \cdot \overline{e_{trg}} = (e_s, e_t) \in \twoheadrightarrow\}$$

$$\mathsf{off}(e, \alpha) = \{e_t \mid e_{trg} \in \mathsf{from}^*(e, \alpha) \wedge \exists e_s \cdot \overline{e_{trg}} = (e_s, e_t) \in \rightarrowtail\}$$

Intuitively from(e_s) returns the hyper edges that start from e_s, from* keeps traversing from to collect all (active) hyper edges triggered from a single edge, and on(e, α) (resp. off(e, α)) collect all the targets triggered from e by an activating (resp. deactivating) edge.

For example, in the vending machine in Fig. 1 we have that off(e_1, E) = $\{e_1, e_2\}$, where $\overline{e_1} = \langle Insert, 1€, Chocolate \rangle$ and $\overline{e_2} = \langle Insert, 0.5€, Coffee \rangle$. This means that executing the edge from *Insert* to *Chocolate* triggers the deactivating edges e_1 and e_2. Using this notion of (de)activation, the evolution of a reactive graph is formalised below.

Definition 3 (Semantics). *The semantics of a reactive graph M is given by the evolution of a configuration $\langle w, \alpha \rangle$ of a state $w \in W$ and active edges $\alpha \subseteq E$, starting from the initial configuration $\langle w_0, \alpha_0 \rangle$, given by the rule below.*

$$\frac{\exists e \in \alpha \cdot \overline{e} = (w, a, w') \wedge \alpha' = \big(\alpha \cup \mathsf{on}(e, \alpha)\big) \backslash \mathsf{off}(e, \alpha)}{\langle w, \alpha \rangle \xrightarrow{a}_M \langle w', \alpha' \rangle}$$

Using the semantics above to our reactive vending machine in Fig. 1 we obtain the LTS depicted in Fig. 2. This semantics differs from Gabbay's semantics by atomically collecting all activate edges before applying their (de)activations effects, instead of activating and deactivating edges during the traversal of triggered edges. It also introduces a bias: whenever an edge is both activated and deactivated in a step, deactivation takes precedence. However this may not be intended, which we will address in the next section over contradictory effects.

Relevant Properties of Reactive Graph. As seen in Definition 3, the behaviour of a reactive graph $M = (W, Act, E, \rightarrow, \twoheadrightarrow, \rightarrowtail, \bar{\cdot}, w_0, \alpha_0)$ from a configuration $\langle w, \alpha \rangle$ can be represented by the LTS induced by relation $\rightarrow_M = \bigcup \{\xrightarrow{a}_M \mid a \in Act\}$. Many standard properties of reactive graphs can be defined over the LTS induced by the semantics of reactive graph, namely:

Deadlocks. A deadlock is a state from which there is no transition (in our case an *active transition*), often undesirable. In reactive graphs we can also search for deadlocks by traversing the induced LTSs from w_0 while searching for states without outgoing transitions.

Unreachable States. An unreachable state, also undesirable in many systems, is a state that cannot be reached from the initial configuration.

Observational Equivalence. As in standard LTS, two configurations are said to be equivalent if they behave in the same way. One way of defining such kind of equivalences is by means of their induced LTS: two configurations are behavioural equivalent if their induced LTS are bisimilar.

Other properties that can be analysed directly over reactive graphs include:

Contradictory Effects. A contradictory effect is when a step triggers both the activation and deactivation of the same edge. The semantics in Definition 3 gives priority to disabling, but often these situations are the result of bad design decisions that should be avoided, and can be signalled as warnings.

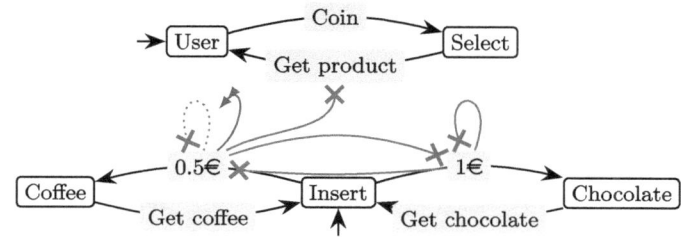

Fig. 3. An example of two reactive graphs and one intrusive edge

Unreachable Transitions. Similarly to unreachable states, (hyper) edges that cannot be fired are usually undesirable or a result of a bad understanding of a system. Hence it is a property that can also be investigated directly over reactive graphs.

Products on Reactive Graphs. Synchronous and asynchronous products of RG can be defined in the standard way. This section discusses a new product, called *intrusive product*. It allows the connections between two RGs, i.e., the execution of an action in a given machine can interfere with the activation/deactivation actions of the other machine, and vice-versa.

Definition 4. *Given two multi-action reactive graphs* M_1, M_2, *and* $\Gamma^\oplus, \Gamma^\ominus \subseteq E_1 \times E_2 \cup E_2 \times E_1$ *is the set of intrusive edges between* M_1 *and* M_2. *The effects produced by* $e \in E_{M_i}$ *in* M_i *is given for the set follow:*

$$\alpha_i(\Gamma^\oplus, \Gamma^\ominus, e) = \big(\alpha_i \cup \mathsf{on}(e, \alpha_i) \cup \Gamma^\oplus(e)\big) \setminus \big(\mathsf{off}(e, \alpha_i) \cup \Gamma^\ominus(e)\big)$$

Figure 3 illustrates an intrusive product of $Usr \parallel_{\varnothing, \Gamma^\ominus} VM$, where Usr is the upper RG and $\Gamma^\ominus = \{\langle\langle Insert, 0.5 €\rangle, Coffee\rangle, \langle User, \mathsf{Get\ product}, Select\rangle\}$.

Formally the asynchronous product ($\parallel_{\Gamma^\oplus, \Gamma^\ominus}$) is defined by the rules below.

$$\frac{\exists\, e \cdot s_1 \xrightarrow{a} s_1' \in \alpha_1 \;\wedge\; \alpha_1' = \big(\alpha_1 \cup \mathsf{on}(e, \alpha_1)\big)\backslash\mathsf{off}(e, \alpha_2) \;\wedge\; \alpha_2' = \alpha_2(\Gamma^\oplus, \Gamma^\ominus, e)}{\langle s_1, \alpha_1\rangle \parallel_{\Gamma^\oplus, \Gamma^\ominus} \langle s_2, \alpha_2\rangle \xrightarrow{a} \langle s_1', \alpha_1'\rangle \parallel_{\Gamma^\oplus, \Gamma^\ominus} \langle s_2, \alpha_2'\rangle}$$

$$\frac{\exists\, e \cdot s_2 \xrightarrow{a} s_2' \in \alpha_2 \;\wedge\; \alpha_2' = \big(\alpha_2 \cup \mathsf{on}(e, \alpha_2)\big)\backslash\mathsf{off}(e, \alpha_2) \;\wedge\; \alpha_1' = \alpha_1(\Gamma^\oplus, \Gamma^\ominus, e)}{\langle s_1, \alpha_1\rangle \parallel_{\Gamma^\oplus, \Gamma^\ominus} \langle s_2, \alpha_2\rangle \xrightarrow{a} \langle s_1, \alpha_1'\rangle \parallel_{\Gamma^\oplus, \Gamma^\ominus} \langle s_2', \alpha_2'\rangle}$$

The synchronous product ($\mathord{\mathscr{S}}_{\Gamma^\oplus, \Gamma^\ominus}$) is based on shared actions, defined below.

$$\frac{\begin{array}{c}\exists\, e_1 \cdot s_1 \xrightarrow{a} s_1' \in \alpha_1 \\ \exists\, e_2 \cdot s_2 \xrightarrow{a} s_2' \in \alpha_2\end{array} \wedge \begin{array}{c}\alpha_1' = \big(\alpha_1 \cup \mathsf{on}(e_1, \alpha_1) \cup \Gamma^\oplus(e_1)\big)\backslash\big(\mathsf{off}(e_1, \alpha_1) \cup \Gamma^\ominus(e_1)\big) \\ \alpha_2' = \big(\alpha_2 \cup \mathsf{on}(e_2, \alpha_2) \cup \Gamma^\oplus(e_2)\big)\backslash\big(\mathsf{off}(e_2, \alpha_2) \cup \Gamma^\ominus(e_2)\big)\end{array}}{\langle s_1, \alpha_1\rangle \, \mathord{\mathscr{S}}_{\Gamma^\oplus, \Gamma^\ominus} \langle s_2, \alpha_2\rangle \xrightarrow{a} \langle s_1', \alpha_1'\rangle \, \mathord{\mathscr{S}}_{\Gamma^\oplus, \Gamma^\ominus} \langle s_2', \alpha_2'\rangle}$$

This product supports the modelling, e.g., of self-adaptive systems with a layer that manages a given system and the actual system being adapted [17].

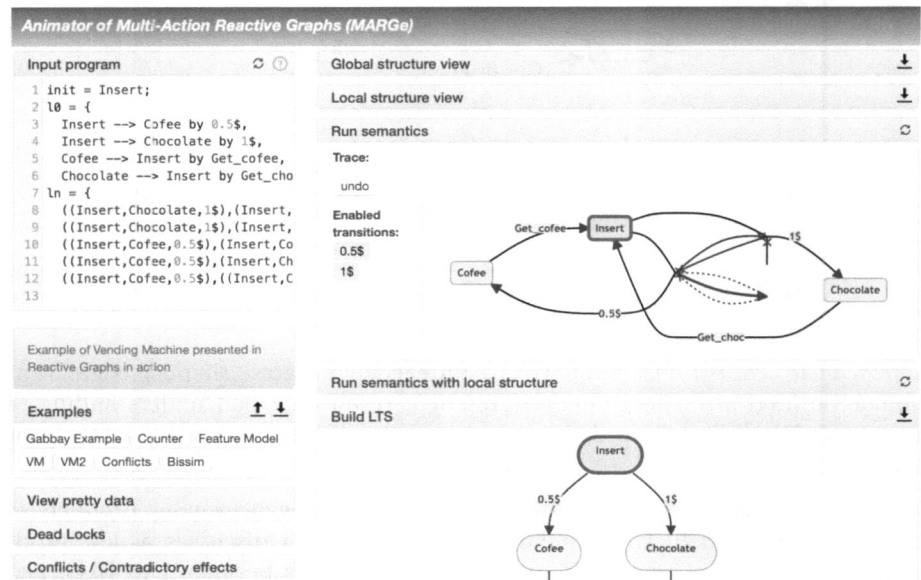

Fig. 4. Screenshot of the web interface of the Marge tool

3   The Marge tool

This section briefly presents the Marge tool and its features. A screenshot of its web interface can be found in Fig. 4 with the vending machine example from Fig. 1. Marge is open-source and developed in *Scala*, using the *Mermaid* library[1] to produces graphical representations of the reactive graph and its semantics. The tool is compiled to *JavaScript* that is used to build an interactive web page, using the CAOS library [18] which includes support to animate operational semantics and compare semantics.

Using the Tool. The tool can be used at https://fm-dcc.github.io/MARGe, illustrated in Fig. 4 with our vending machine example. The model is introduced using a textual description in the "Input program" widget (top left). It can also be found in the list of examples (middle left). The remaining widgets provide our analysis and visualisations, and can be either collapsed (as the "Global structure view") or expanded (as the "Run semanitcs").

Available Widgets. Some of the available widgets are described below.

- Input program – uses a textual notation, mimicking the mathematical structures, not yet optimised to be compact and maintainable.
- Global structure view – shows the graphical representation of a reactive graph. A simplified version without hyper edges and deactivated edges is depicted in the widget "local structure view".

---

[1] Mermaid is popular markup language for diagrams, cf. https://mermaid.js.org.

- **Run semantics** – allows the user to simulate the reactive graph by selecting, at each step, an active transition that should be taken. After selecting this transition the graph is updated, including the active edges and the current state.
- **Genereted LTS** – displays the underlying LTS by expanding all possible actions of the the reactive graph (up to a fixed bound).
- **Number of states and edges** – presents the number of states and edges of both the reactive graph and its encoded LTS. E.g., our vending machine as a similar number of states and edges. But a variation (available online) with a limited stock (instead of limited money) uses 4 states, 5 ground edges, and 3 hyper edges, against 19 states and 20 edges in the encoded LTS.
- **Find strong bisimulation** – checks if two reactive graphs separated by '∼' in the input program are equivalent (i.e., bisimilar), providing either a bisimulation or an an explanation for not finding one.
- **Conflicts/Contradictory effects** – finds conflicts when they exist, i.e. traces until a transition that simultaneously tries to activate and deactivate.
- **DeadLocks** – checks the existence of deadlocks in the behaviour of a given reactive graph.
- **Products** – this a set of widgets, presenting the different types of product introduced in this paper.

4   Conclusions and Future Work

Reactive graphs can provide a compact and insightful representation of a variety of reconfigurable scenarios, e.g., in the context of communication protocols [11] and in a biological setting [20]. They are also closely related to van Benthem's game models with adversarial agents [4] and to Areces et al.'s logic with reconfiguring modal operators [1]. Extensions to reactive graphs with the paraconsistency paradigm have also been recently proposed [9]. Most work on reactive graphs is theoretical, and a small effort has been done to provide tool support and automatization of results. This paper presents the tool Marge with basic editor and exploration mechanisms of reactive graphs, including specific analysis such as a search for contradictory effects. As future work, we intent to improve the usability of Marge (e.g., improving the input language), extend it to support fuzzy extensions (to measure, e.g., costs and rewards from applying reconfigurations [6,20]), and to integrate a model checker according to a suitable logic. The latter would be similar to how we integrated the mCRL2 toolset [5] to analyse connectors [19] and team automata [2], based on a predecessor of CAOS [10].

Acknowledgments. This work is supported by the FCT, the Portuguese funding agency for Science and Technology, with the projects UIDB/04106/2020 (https://doi.org/10.54499/UIDB/04106/2020), UIDP/04106/2020 (https://doi.org/10.54499/UIDP/04106/2020) and PTDC/CCI-COM/4280/2021. It was also supported by the CISTER Research Unit (UIDP/UIDB/04234/2020), financed by National Funds through FCT/MCTES and by project Ibex (ref. PTDC/CCI-COM/4280/2021) financed by national funds through FCT.

References

1. Areces, C., Fervari, R., Hoffmann, G.: Relation-changing modal operators. Log. J. IGPL **23**(4), 601–627 (2015). https://doi.org/10.1093/JIGPAL/JZV020
2. ter Beek, M.H., Cledou, G., Hennicker, R., Proença, J.: Can we communicate? using dynamic logic to verify team automata. In: Chechik, M., Katoen, J.P., Leucker, M. (eds.) FM 2023. LNCS, vol. 14000. Springer, Heidelberg (2023). https://doi.org/10.1007/978-3-031-27481-7_9
3. ter Beek, M.H., Damiani, F., Lienhardt, M., Mazzanti, F., Paolini, L.: Efficient static analysis and verification of featured transition systems. Empir. Softw. Eng. **27**(1), 10 (2021). https://doi.org/10.1007/s10664-020-09930-8
4. van Benthem, J.: An essay on sabotage and obstruction. In: Hutter, D., Stephan, W. (eds.) Mechanizing Mathematical Reasoning, Essays in Honor of Jörg H. Siekmann on the Occasion of His 60th Birthday. LNCS, vol. 2605, pp. 268–276. Springer, Heidelberg (2005). https://doi.org/10.1007/978-3-540-32254-2_16
5. Bunte, O., et al.: The mCRL2 toolset for analysing concurrent systems. In: Vojnar, T., Zhang, L. (eds.) TACAS. LNCS, vol. 11428, pp. 21–39. Springer, Heidelberg (2019). https://doi.org/10.1007/978-3-030-17465-1_2
6. Campos, S., Santiago, R.H.N., Martins, M.A., Figueiredo, D.: Introduction to reversal fuzzy switch graph. Sci. Comput. Program. **216**, 102776 (2022). https://doi.org/10.1016/J.SCICO.2022.102776
7. Capilla, R., Bosch, J., Trinidad, P., Ruiz-Cortés, A., Hinchey, M.: An overview of dynamic software product line architectures and techniques: observations from research and industry. J. Syst. Softw. **91**, 3–23 (2014). https://doi.org/10.1016/j.jss.2013.12.038
8. Cordy, M., Classen, A., Heymans, P., Legay, A., Schobbens, P.Y.: Model checking adaptive software with featured transition systems. In: Assurances for Self-Adaptive Systems: Principles, Models, and Techniques, pp. 1–29 (2013)
9. Costa, D., Figueiredo, D., Martins, M.A.: Relation-changing models meet para-consistency. J. Log. Algebraic Methods Program. **133**, 100870 (2023). https://doi.org/10.1016/J.JLAMP.2023.100870
10. Cruz, R., Proença, J.: Reolive: Analysing connectors in your browser. In: Mazzara, M., Ober, I., Salaün, G. (eds.) STAF 2018. LNCS, vol. 11176, pp. 336–350. Springer, Heidelberg (2018). https://doi.org/10.1007/978-3-030-04771-9_25
11. Figueiredo, D., Martins, M.A., Barbosa, L.S.: A note on reactive transitions and reo connectors. In: de Boer, F.S., Bonsangue, M.M., Rutten, J. (eds.) It's All About Coordination - Essays to Celebrate the Lifelong Scientific Achievements of Farhad Arbab. LNCS, vol. 10865, pp. 57–67. Springer, Heidelberg (2018). https://doi.org/10.1007/978-3-319-90089-6_4
12. Gabbay, D.M.: Reactive Kripke models and contrary to duty obligations. Part A: semantics. J. Appl. Logic **11**(1), 103–136 (2013). https://doi.org/10.1016/j.jal.2012.08.001
13. Gabbay, D.M.: Reactive Kripke Semantics. Cognitive Technologies, Springer, Heidelberg (2013). https://doi.org/10.1007/978-3-642-41389-6
14. Gabbay, D.M., Marcelino, S.: Modal logics of reactive frames. Stud. Logica. **93**(2–3), 405–446 (2009). https://doi.org/10.1007/S11225-009-9214-1
15. Gabbay, D.M., Marcelino, S.: Global view on reactivity: switch graphs and their logics. Ann. Math. Artif. Intell. **66**(1–4), 131–162 (2012). https://doi.org/10.1007/S10472-012-9316-8

16. Harel, D., Pnueli, A.: On the development of reactive systems. In: Apt, K.R. (ed.) Logics and Models of Concurrent Systems, pp. 477–498. Springer, Heidelberg (1985). https://doi.org/10.1007/978-3-642-82453-1_17
17. Päßler, J., ter Beek, M.H., Damiani, F., Tarifa, S.L.T., Johnsen, E.B.: Formal modelling and analysis of a self-adaptive robotic system. In: Herber, P., Wijs, A. (eds.) iFM 2023. LNCS, vol. 14300, pp. 343–363. Springer, Heidelberg (2023). https://doi.org/10.1007/978-3-031-47705-8_18
18. Proença, J., Edixhoven, L.: Caos: a reusable scala web animator of operational semantics. In: Jongmans, S.S., Lopes, A. (eds.) COORDINATION 2023, DisCoTec 2023. LNCS, vol. 13908, pp. 163–171. Springer, Heidelberg (2023). https://doi.org/10.1007/978-3-031-35361-1_9
19. Proença, J., Madeira, A.: Taming hierarchical connectors. In: Hojjat, H., Massink, M. (eds.) FSEN 2019. LNCS, vol. 11761, pp. 186–193. Springer, Heidelberg (2019). https://doi.org/10.1007/978-3-030-31517-7_13
20. Santiago, R.H.N., Martins, M.A., Figueiredo, D.: Introducing fuzzy reactive graphs: a simple application on biology. Soft. Comput. 25(9), 6759–6774 (2021). https://doi.org/10.1007/S00500-020-05353-1
21. Tinoco, D., Madeira, A., Martins, M.A., Proença, J.: Reactive graphs in action (extended version). CoRR (2024). https://doi.org/10.48550/arXiv.2407.14705

Security and Blockchain

Extracting Formal Smart-Contract Specifications from Natural Language with LLMs

Gabriel Leite[1]([✉]), Filipe Arruda[1][iD], Pedro Antonino[2][iD], Augusto Sampaio[1][iD], and A. W. Roscoe[2,3][iD]

[1] Centro de Informática, Universidade Federal de Pernambuco, Recife, Brazil
{gnl2,fmca,acas}@cin.ufpe.br
[2] The Blockhouse Technology Limited, Oxford, UK
pedro@tbtl.com
[3] University College Oxford Blockchain Research Centre, Oxford, UK

Abstract. Developers tend to be reluctant to provide formal specifications for software components; even well-established design-by-contract (DbC) properties like invariants, pre- and postconditions are neglected. This has hindered a more widely practical dissemination of the DbC paradigm. In this paper, we employ state-of-the-art NL processing technologies, using Large Language Models (LLMs), particularly, ChatGPT, to automatically infer formal specifications from component textual behavioural descriptions. More specifically, we implemented a framework (DbC-GPT), parameterised by a context, which is able to generate postcondition specifications for smart contract functions implemented in Solidity. The output of DbC-GPT is in the notation of the solc-verify tool (a verifier for Solidity) that is used to: (i) check the syntax of the inferred specification; and (ii) verify whether a reference implementation conforms to this specification. This is carried out in a loop in such a way that the DbC-GPT context is iteratively improved with verification counterexamples. To evaluate DbC-GPT, we have used some Ethereum standards (ERC20, ERC721, and ERC1155) and compared the precision of the generated specifications for several GPT contexts that consider information of these standards in isolation as well as their combination.

Keywords: smart contracts · natural language processing · design-by-contract · formal verification · LLMs · GPT · ChatGPT

1 Introduction

The development of smart contracts is marked by their immutable and autonomous nature, which demands high precision in their specification and implementation. While there are several approaches to specify and reason about smart contracts [5, 12, 15, 17, 26], the widespread integration of formal methods into software development processes continues to be a significant challenge.

D. Marmsoler and M. Sun (Eds.): FACS 2024, LNCS 15189, pp. 109–126, 2024.
https://doi.org/10.1007/978-3-031-71261-6_7

An approach that has gained progressively more attention to tackle this issue is *hidden formal methods*, whose goal is to make formal methods more accessible and practical by integrating them seamlessly, and in a way as transparent as possible, into existing development workflows. The usual starting point of such approaches are requirements expressed in natural language (NL). By using NL processing (NLP) techniques, these approaches transform NL requirements into more precise models that can then be used for code generation [10,11], formal verification [7,9], testing [6,16,25], and so on.

Large Language models (LLMs), including GPT (Generative Pre-trained Transformer) variants, utilise deep learning to capture the nuances of natural language, making them particularly suited for tasks that require a deep semantic understanding of text, significantly reducing the risk of human error and misinterpretation [24]. Furthermore, these AI-driven approaches support continuous learning and adaptation, which allows them to improve over time as they are exposed to more examples and potential corrections [8]. Some works have recently used LLMs to automate the extraction and processing of text for several different purposes, including the generation of formal specifications from requirements documents [14]. Some other approaches have used LLMs to infer specification elements directly from implementations, such as the generation of loop invariants from C programs [13].

Despite the several efforts in this direction, the overall question of to what extent LLMs can systematically support formal specification and reasoning in software development needs to be more deeply investigated.

The main purpose of this paper is to integrate these transformer models (specifically ChatGPT [8]) into a framework, that we call DbC-GPT, which can automate the extraction of formal specifications for smart contracts from natural language descriptions. Although this can be generally applied to different languages, we focus on Solidity [1]. Specifically, DbC-GPT generates specifications in the design-by-contract paradigm, with focus on postconditions for each function of a Solidity smart contract. Preconditions and postconditions share the same nature as they both serve as formal specifications defining the expected properties or constraints at different points of function execution-before and after, respectively. We consider only postconditions because we reason about contracts as open programs since there is hardly any control concerning the caller of a contract deployed in a blockchain, and a precondition imposes a responsibility to be obeyed by the caller, which, in this context, is not generally possible to verify. We consider invariants as a topic for future work.

By focusing typically on coding, developers often hesitate to supply formal specifications, hindering a broader practical adoption of the design-by-contract paradigm. Furthermore, even when specifications are provided, an automated framework, as the one proposed here, helps to mitigate risks associated with manual encoding of these specifications from requirements, which tends to be error-prone. In the particular context of our research to build a fully automatic trusted deployer for smart contracts running in blockchains [3,4], DbC-GPT fills an important gap since it automatically generates the specifications that are fed into the trusted deployer to ensure a safe deploy and subsequent evolution of smart contract implementations.

The DbC-GPT framework takes three inputs: an Ethereum Improvement Proposal (EIP), an interface (with the functions to be specified), and a reference Solidity implementation for this standard. Its output is a design-by-contract specification in the notation of the solc-verify tool (a verifier for Solidity); as a constraint, the reference implementation provided as input must conform to the produced specification. Conformance is the usual (partial) correctness notion adopted in DbC approaches: assuming the precondition (in our context, this is always the true predicate) of a function holds, its execution must obey the respective postcondition, provided the execution terminates. The framework is parameterised by a (LLM) model and a demonstration context. In this paper, we use ChatGPT's GPT-4o model and contexts, but we could have instantiated our framework with Llama [23] instead, for instance.

For evaluation purposes, we have explored 8 demonstration contexts, which are differentiated in the way we provide information about the Ethereum standards. These examples contain example reference specifications of these standards so that the ChatGPT model has instances of what type of specification it should generate. We use solc-verify to check for conformance of the reference implementation with respect to the generated specification. This verifier is also used to check the syntax of the inferred specification. For all these contexts, verification is used in a feedback loop so that counterexamples (or verification errors, in general) are used to extend the respective learning context. We then compared these instances of the framework, each instantiated with one of the GPT contexts, concerning the precision of the generated specifications.

A distinguishing feature of our approach is the verification loop in the process. What we propose in this paper is a sort of counterexample-guided specification generation framework. We show that this provides relevant feedback information to improve the context so that progressively more precise specifications are generated by DbC-GPT. This design is based on the idea of *in-context learning* [8], namely, that GPT models perform well when the learning context is extended to refine the scope of the task being solved.

The next section provides some background on Solidity and solc-verify based on a small fragment of the Ethereum ERC20 standard that we also use as a running example. Section 3 provides an overview of the proposed framework and details of its design as a generative specification mechanism parameterised by customised GPT (demonstration) contexts. The evaluation of the framework with some Ethereum standards is presented in Sect. 4. Related work is addressed in Sect. 5. We then conclude with a summary of the contributions, limitations, and topics for future work.

2   Background

The ERC20 is a standard interface for fungible tokens on the Ethereum blockchain. It includes several key functions, such as transferring tokens, approving allowances, and transferring tokens on behalf of others. The description of an Ethereum standard is given by a document called Ethereum Improvement Proposals (EIP) such as, for example, that for the ERC20[1]. This includes the standard interface for each function, and an associated informal specification. An example is given for the ERC20 function `transfer` in Fig. 1. This function has two input parameters (`_to` for the receiver address and `_value` for the amount to be transferred); as explained in the sequel, the address for the sender is an implicit parameter in Solidity. There is also a boolean output parameter (`success`) used to record whether the function execution is successful (`true`) or not (`false`). The textual specification of the function is self-explanatory.

transfer

Transfers `_value` amount of tokens to address `_to`, and MUST fire the `Transfer` event. The function SHOULD `throw` if the message caller's account balance does not have enough tokens to spend.

Note Transfers of 0 values MUST be treated as normal transfers and fire the `Transfer` event.

```
function transfer(address _to, uint256 _value) public returns (bool success)
```

Fig. 1. EIP informal specification for the `transfer` function

Concerning a reference implementation, our running example is based on the OpenZeppelin[2] of the ERC20 standard. The Solidity code snippet for the transfer function is presented in Listing 1.1.

```
1  contract ERC20 is IERC20 {
2    /* [...] */
3    mapping (address => uint256) private _balances;
4
5    function transfer(address to, uint256 amount) public virtual override
          returns (bool) {
6      address owner = _msgSender();
7      _transfer(owner, to, amount);
8      return true;
9    }
10   function _transfer(address sender, address recipient, uint256 amount)
          internal {
11     require(sender != address(0), "ERC20: transfer from the zero address
          ");
12     require(recipient != address(0), "ERC20: transfer to the zero
          address");
13
14     _balances[sender] = _balances[sender].sub(amount, "ERC20: transfer
          amount exceeds balance");
15     _balances[recipient] = _balances[recipient].add(amount);
16     emit Transfer(sender, recipient, amount);
17   }
18 }
```

Listing 1.1. OpenZeppelin implementation for the transfer function

---

[1] https://eips.ethereum.org/EIPS/eip-20.

[2] https://raw.githubusercontent.com/OpenZeppelin/openzeppelincontracts/ 19c74140523e9af5a8489fe484456ca2adc87484/contracts/token/ERC20/ERC20. sol.

A contract in Solidity allows the declaration of types, attributes and functions. In our example fragment, we show only one attribute, `_balances` (Line 3): a mapping from addresses (represented by a 160-bit number) to token balances represented by unsigned (256-bit) integers. The function `transfer` (Line 5) transfers a token amount (parameter `amount`) from the caller (whose address is stored in an implicit argument `msg.sender`) to a destination address (parameter `_to`). This function yields a boolean value that states whether the execution was successful. Note that, despite the differences in some parameter names, the signature of this function is clearly the same as that in the fragment ERC20 EIP.

In the first line of the function implementation, the caller address `msg.sender` is yielded by the (private, internal) function `_msgSender()` that is imported from another Solidity contract; its definition is straightforward and omitted here. This address is saved in the local variable `owner`. Next, `transfer` delegates the actual implementation via a call to the internal function `_transfer` that has an additional parameter to those of `transfer` to record the address of the caller. Finally, the `true` value is returned to indicate the function's successful execution.

The applicability of a function can be captured using the `require(condition)` statement. If the `condition` holds, the execution proceeds normally; otherwise, the function execution aborts and the state before the start of the function execution is preserved. The two `require` clauses of `_transfer` impose that the addresses of both the sender and of the recipient must not be the zero address. These clauses include respective error messages. The next two assignments capture the effect of the transfer: the balance of the sender is decreased by `amount` and that of the destination is increased by the same amount. These statements make use of the functions `sub` and `add`, for integer subtraction and addition; `sub` throws an error message if the result is less than zero, and `add` avoids integer overflow by limiting the sum to the highest integer (2^{256}). The `emit` keyword is used to communicate an event in Solidity; in our example, it is used to log the transfer transaction.

A formal DbC specification in the solc-verify notation for the function `transfer` is given in Listing 1.2.

```
1  /// @notice postcondition (_balances[msg.sender] == __verifier_old_uint(
       _balances[msg.sender]) - amount && msg.sender != to ) || (_balances[
       msg.sender] ==  __verifier_old_uint(_balances[msg.sender]) && msg.
       sender == to)
2
3  /// @notice postcondition (_balances[to] == __verifier_old_uint(
       _balances[to]) + amount  && msg.sender != to ) || (_balances[to] ==
       __verifier_old_uint(_balances[to] ) && msg.sender == to)
```

Listing 1.2. Postconditions for the `transfer` function using the solc-verify syntax

It includes two postconditions (`@notice postcondition`) that are implicitly conjoined. The first postcondition determines that the sender's balance should be decreased by the transferred amount, provided that the sender is not the recipient; otherwise, the balance must be preserved. For an attribute x, the expression `__verifier_old_uint(x)` holds the value of `x` at the start of the function execution. The notation for conjunction (`&&`), disjunction (`||`) and negation (`!`) are standard. The second postcondition states that the recipient's balance

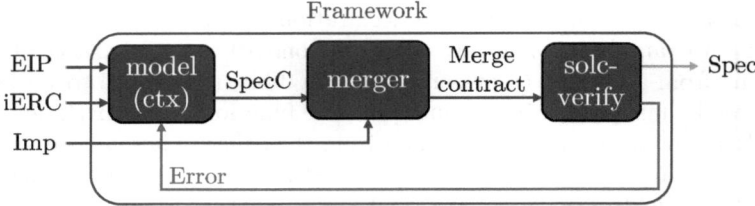

Fig. 2. Framework architecture overview.

must increase by the transferred amount, unless the sender is also the recipient, in which case the balance must be preserved.

In the following sections, we use this running example to illustrate the application of our DbC-GPT framework for generating formal specifications and verifying them using solc-verify.

3   Framework

Our framework's architecture is given in Fig. 2. It takes as input a specification of an Ethereum standard in the form of a textual EIP (represented by EIP), an annotated ERC interface (iERC), and a reference implementation in the form of a Solidity smart contract, represented by Imp. Given these inputs, the framework generates a DbC specification denoted by Spec.

Internally, the framework is parameterised by an LLM model and a *demonstration context*; the latter consists of some examples of output (i.e. reference specifications) that should help the model in finding the appropriate specification. These elements are combined with a smart contract verifier (solc-verify) in a *counterexample-guided generation process*. The candidate specification (SpecC) generated (given the model and learning context defined) is checked for the syntactic and semantic validity by solc-verify. Semantic validity concerns whether the given reference implementation conforms to the specification in the usual DbC context, but restricted to partial correctness. If solc-verify does not show conformance, the error witnessed is returned to the model (or rather added to the learning context) that generates a new candidate specification. This loop continues until a valid specification is generated or a pre-determined number of iterations is reached. If a valid specification is generated, it is returned to the user of the framework. In the following, we detail this process by describing our methodology to create the *main prompt* (which encompasses the demonstration context) passed to our model, and how our loop is implemented. The model we use in this paper is ChatGPT's GPT-4o[3].

Our framework relies on a property of LLMs, such as ChatGPT, called *in-context learning* (sometimes also referred to as few-shot learning) [8], that is, such models have been shown to perform well when they are given a *learning context*

---

[3] https://openai.com/index/hello-gpt-4o/.

that guides their task-solving process. For instance, given a few examples (i.e. demonstrations) on how the task can be solved for other instances before asking for the solution of the specific instance desired. A concrete example that is often used is to provide many translations of words from one language to another, before asking for the translation of the specific word of interest [8]. We point out that this learning context provided is not part of the model's training (i.e. it does not affect its internal weights) instead it helps the (pre-trained) model to more precisely navigate its "search space" when looking for the correct solution to the task. Both our demonstration context and our counterexample-guided process are designed to make use of this property.

3.1   Main Prompt Preparation

The main prompt is the starting point of the learning context in our framework, namely, it alone represents the initial learning context given to the model to generate the first candidate specification. The preparation of this prompt involved an ad-hoc *prompt engineering* process[4] where we iteratively refined prompts until we reached the version we present here. In this section, we also illustrate some lessons learned in this prompt engineering process.

The main prompt is formed of three sections: (i) a fixed general description of (i.e. the framing) the task, (ii) a number of examples of outputs that should guide the model in finding the appropriate specification (i.e. *the demonstration context*), and (iii) the inputs (EIP and annotated ERC interface) for the specific generation task we want the model to accomplish.

The first (framing) section is given by Listing 1.3. In the first paragraph, the prompt generically captures the task of generating a formal specification from an ERC interface and an EIP textual file. Note that we point out that only postconditions are to be generated. We also add some instructions and guidance to help the model find the right kind of specification. We instruct the model not to generate function implementations (Line 5), and what is the syntax of a solc-verify postcondition (Line 6) and its positioning with respect to the function for which it is expected to generate the specification (Line 7). Moreover, we have also added some guidance on the generation of the postconditions; they provide some hints as to what semantic aspects the postcondition should capture. In our initial attempts to create a prompt for this task, the ChatGPT model would generate a specification with function implementations or preconditions. Both of these things are not allowed in our specification so we had to refine our initial prompt to reflect these requirements.

The second section of our main prompt is the demonstration context; it gives examples of (i.e. demonstrates) reference specifications for ERCs. This section is parameterised by a (possibly empty) sequence of ERCs. They represent demonstrations of expected outputs (i.e. valid specifications) that should help guide the GPT model in its search process. The reference specification of each ERC in this sequence is added, in the order given, as illustrated in Listing 1.4; it exemplifies

---

[4] https://platform.openai.com/docs/guides/prompt-engineering.

```
1  Given an examaple of ERC interface, the ERC interface to be annotated
        and an EIP markdown, generate a specification for the ERC interface
        with solc-verify postconditions annotations, just postconditions, no
        other annotations types, this is very important!
2
3      Instructions:
4
5      - Function Bodies: The specification must not contain function
            implementations.
6      - Postconditions Limit: Each function must have at most 4
            postcondition (/// @notice postcondition) annotations above
            the function signature. Do not exceed this limit under any
            circumstances.
7      - Position: add the solc-verify annotation above the related
            function, example:
8          /// @notice postcondition supply == _totalSupply
9          function totalSupply() public view returns (uint256 supply);
10     - Output format: return the annotated interface inside code
            fence (''') to show the code block. RETURN JUST THE CONTRACT
            ANNOTATED, NOTHING MORE.
11
12     Guidance for Generating Postconditions:
13
14     - State Changes: Reflect how state variables change. For example
            , ownership transfer should reflect changes in token
            ownership and balances.
15     - Conditions on Input: Consider how inputs affect the state
            variables.
16     - Reset Conditions: Ensure certain variables are reset after the
            function execution, if applicable.
```

Listing 1.3. General task description section.

the case of the sequence ⟨ERCA, ERCB⟩ where $reference_specification_ERCA$ and $reference_specification_ERCB$ are placeholders for where the reference specifications of these ERCs would be placed. For the empty sequence, this section of the prompt is omitted. Again, the addition of these demonstrations follows the principle that GPT models are few-shot learners. During our prompt engineering process, we have observed that passing EIPs in addition to reference specifications in this demonstration section was not helpful as it was hindering the specification search process.

```
1      ERC interface example:
2      '''solidity
3          $reference_specification_ERCA$
4      '''
5
6      ERC interface example:
7      '''solidity
8          $reference_specification_ERCB$
9      '''
```

Listing 1.4. Demonstration context section for ERC sequence ⟨ERCA, ERCB⟩ .

The last section defines the inputs for the specification generation process, namely, it instantiates the specification generation task. Of course, this part also varies. Listing 1.5 illustrates how this section is constructed for a given ERC; the placeholder $ERC_annotated_interface$ (Line 6) denotes where the annotated interface should be placed, EIP_name (Line 9) where its name should go, and EIP_text (Line 12) where its EIP textual description should be placed. Line 3 is added only if the demonstration section is non-empty.

```
1       Can you please generate a specification given the following ERC
           interface   (delimited by token ''' solidity ''' ) and EIP
           markdown (delimited by token <eip>)?
2
3       HERE FOLLOWS THE CONTRACT TO ADD SOLC-VERIFY ANNOTATIONS, LIKE
           THE EXAMPLES ABOVE:
4
5       '''solidity
6       $ERC_annotated_interface$
7       '''
8
9       EIP $EIP_name$ markdown below:
10
11      <eip>
12      $EIP_text$
13      </eip>
```

Listing 1.5. ERC input section of main prompt.

In engineering this last section of the prompt, we came across the most interesting prompt refinement. The model was much more sensitive to the EIP's textual specification when it was (additionally) used to annotate the ERC interface. Thus, for each function in the interface, we have extracted the parts of the EIP which are related to it and added as a comment to this function in the ERC interface. For instance, Listing 1.6 illustrates how the **transfer** function is annotated with its corresponding EIP specification. For the specifications that we are interested in this paper, identifying these EIP extracts is fairly simple given that EIPs are already structured in a similar (per function) way. We conjecture that the verbosity of the EIPs might hinder the models ability to identify the text related to a function's specification.

```
1           /**
2            * Transfers '_value' amount of tokens to address '_to',
                 and MUST fire the 'Transfer' event.
3            * The function SHOULD 'throw' if the message caller's
                 account balance does not have enough tokens to spend
                     .
4
5            * *Note* Transfers of 0 values MUST be treated as normal
                 transfers and fire the 'Transfer' event.
6            */
7           $ADD POSTCONDITION HERE
8           function transfer(address to, uint value) public returns
                 (bool success);
```

Listing 1.6. Example of annotated ERC20 interface for function **transfer**.

3.2   Counterexample-Guided Specification Generation

Once the learning context has been initialised with the main prompt, our framework relies on a counterexample-guided process for the generation of the specification. If the candidate specification generated by our model is not valid, we return the error information provided by the verifier to the model. More precisely, the error is used to extend and enrich the learning context. In this way, we refine the search for a valid specification.

The specification output by the model consists of a smart contract with function signatures and their corresponding specifications in the solc-verify notation. For checking its validity, we generate a Solidity contract that combines the candidate specification with the reference implementation; we call it a *merge* contract [4]. This combination is carried out by our *merger* component as per Fig. 2. Listing 1.7 illustrates an extract of a merge contract where we have combined the implementation (in Listing 1.1) and the specification (in Listing 1.2) of the function `tranfer`. The merge contract brings the annotations that are part of the candidate specification with the function bodies, and auxiliary code, in the implementation. The merge contract (together with auxiliary implementation code) is then checked by solc-verify.

```
1  /// @notice postcondition (_balances[msg.sender] == __verifier_old_uint(
       _balances[msg.sender]) - amount && msg.sender != to ) || (_balances[
       msg.sender] ==  __verifier_old_uint(_balances[msg.sender]) && msg.
       sender == to)
2  /// @notice postcondition (_balances[to] == __verifier_old_uint(
       _balances[to]) + amount  && msg.sender  != to ) || (_balances[to] ==
       __verifier_old_uint(_balances[to] ) && msg.sender == to)
3  function transfer(address to, uint256 amount) public virtual override
       returns (bool) {
4      address owner = _msgSender();
5      _transfer(owner, to, amount);
6      return true;
7  }
```

Listing 1.7. Extract of merge contract with the transfer function.

This verifier provides a report which might indicate a syntactic or semantic problem with the implementation, or that it satisfies the specification. In the latter case, the candidate specification has been proved valid and so it is output by the framework. In the former case, we append the output of solc-verify to the *reinforcing* prompt given in Listing 1.8. This prompt simply restates some of the instructions given in the initial prompt—once more, these extra instructions were added due to prompt engineering.

```
1  Instructions:
2      - Function Bodies: The specification must not contain function
          implementations.
3      - Postconditions Limit: Each function must have at most 4
          postcondition (/// @notice postcondition) annotations above the
          function signature. Do not exceed this limit under any
          circumstances.
4      - Position: add the solc-verify annotation above the related
          function, example:
5      /// @notice postcondition supply == _totalSupply
6      function totalSupply() public view returns (uint256 supply);
7      - Output format: return the annotated interface inside code fence
          (''') to show the code block. RETURN JUST THE CONTRACT ANNOTATED
          , NOTHING MORE.
```

Listing 1.8. Reinforcement prompt passed back to model.

Listing 1.9 illustrates the kind of output of the solc-verify tool can generate. It presents a syntactical error in the specification generated for the function `totalSupply`: the model has generated a postcondition that uses the variable `supply` but this variable does not represent either a member variable in the contract or a parameter of the function.

```
1  Error while running compiler, details: Warning: This is a pre-release
       compiler version, please do not use it in production.
2
3  ======= Converting to Boogie IVL =======
4
5  ======= ./solc_verify_generator/ERC20/imp/ERC20_merge.sol =======
       Annotation:1:1: solc-verify error: Undeclared identifier. supply ==
       _totalSupply ^----^
6
7  ======= ./solc_verify_generator/ERC20/imp/IERC20.sol =======
8
9  ======= ./solc_verify_generator/ERC20/imp/math/SafeMath.sol ======= ./
       solc_verify_generator/ERC20/imp/ERC20_merge.sol:38:5: solc-verify
       error: Error(s) while translating annotation for node function
       totalSupply() public view returns (uint256) { ^ (Relevant source
       part starts here and spans across multiple lines).
```

Listing 1.9. Extract of counterexample passed back to ChatGPT.

The prompt combining the reinforcement of instructions with the output of solc-verify extends the learning context provided to the model in the search of a specification; the invalid specification that triggered the creation of this prompt is also part of this context. The next iteration of our loop uses the extended context to generate a new specification, and this process repeats until a parameterised number of iterations is reached. At that point, our framework stops trying to generate a specification.

It is worth mentioning that this counterexample-guided process based on an implementation biases, to some extent, the search for a specification. Ideally, one would look for the most general (weakest) specification capturing the formal properties for a given contract interface and the respective textual specification. However, when the search is driven (even considering the input textual specification) by an implementation, there is a risk that the specification found may be too narrow and applicable only to that single implementation.

Another risk of performing such a kind of search is that the reference implementation may be wrong. In this case, it may drive the generation process away from the textual specification and into a wrong formal specification. For the time being, we do not consider these cases and leave them for future work.

4   Evaluation

For the evaluation of our framework, we consider ERCs 20, 721, 1155. For each of them, we have created a reference specification, and have copied its EIP, its interface, and a popular reference implementation. All these artefacts together with the source code for our tool and scripts to run our evaluation are available in our repository[5]. For our evaluation section, we parameterise our framework

---
[5] https://github.com/formalblocks/DbC-GPT.

Table 1. Evaluation results. For each demonstration context and input ERC, we describe how many times out of the 10 runs a valid specification was generated and the interval on the number of iterations necessary to generate this many specifications.

Dem. Context	Input		
	20	721	1155
ϵ	2 [2;4]	1 [9;9]	0
20	10 [0;3]	5 [1;4]	✓ 2 [8;9]
721	4 [1;5]	10 [0;0]	✓ 2 [6;9]
1155	5 [1;5]	6 [1;4]	✓ 10 [0;2]
20, 721	10 [0;7]	10 [0;1]	✓ 3 [7;9]
20, 1155	10 [0;5]	7 [1;2]	✓ 10 [0;0]
721, 1155	7 [1;7]	10 [0;4]	✓ 10 [0;0]
20, 721, 1155	9 [0;4]	10 [0;4]	✓ 10 [0;2]

with 8 different demonstration contexts. As per Sect. 3.1, we use sequences of ERCs to represent the different demonstration contexts that we evaluated. We evaluate all the contexts originating from subsequences of $\langle 20, 721, 1155 \rangle$. The empty sequence is denoted by ϵ.

An evaluation step consists of assessing a demonstration context with an input ERC—the latter means its EIP, annotated ERC interface, and reference implementation. We have evaluated each demonstration context with each input ERC leading to 24 evaluation steps, and we repeat each of these steps 10 times. The repetitions accounts for the fact that LLMs are stochastic models so they might output different results even when presented with the same (input) learning context. Each repetition is a different run of the framework, namely, they start with a fresh learning context, and so there is no "learning" (i.e. context/state sharing) between these runs/repetitions. For this evaluation, we parameterised our framework to try to generate a specification with at most 10 loop iterations. These results are summarised in Table 1. In this table, each cell summarises each of these 10 repetitions by outlining for how many of these repetitions a valid specification was found and, in brackets, the interval on the numbers of iterations necessary to find these specifications. For instance, the cell for demonstration context *20, 1155* and input *721* depicts that out of the 10 repetitions 7 were able to find a valid specification and these runs have taken between 1 and 2 counterexample-guided loop iterations. We use 0 iterations to represent that the framework found a specification with the main prompt alone, that is, without needing even one iteration of our counterexample-guided loop.

Our results show how the demonstration context indeed helps to improve the effectiveness of the generation process. Note, for instance, the significant differences between the runs with an empty demonstration context and the ones with non-empty ones. We point out as well that the intricacies of the specification for the ERC1155 token standard translates into our framework's low effectiveness

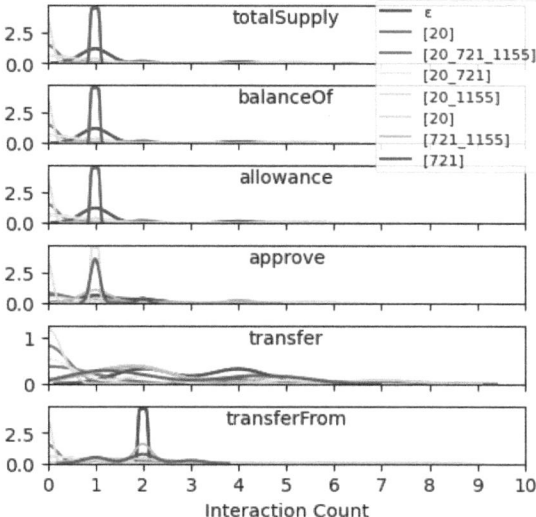

Fig. 3. Per-function analysis for ERC20.

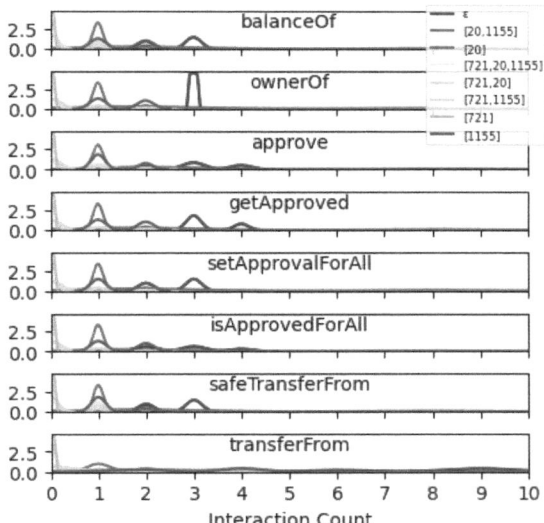

Fig. 4. Per-function analysis for ERC721.

in generating its specification when compared to the other two standards. This specification is expected to use quantification while the others are not. Finally, note that our results provide evidence for the efficacy of our loop strategy. For the majority of the runs, the framework needed counterexample-guided loop iterations to find a valid specification.

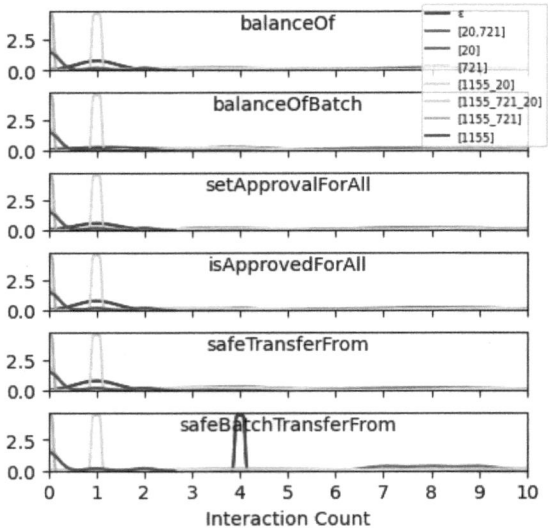

Fig. 5. Per-function analysis for ERC1155.

Note that we have carried out evaluation steps where we try to generate a specification for an (input) ERC that for which the reference specification is being passed in the demonstration context. This may sound like an odd step to analyse but we found it compelling given LLM's stochastic nature and for the purpose of sanity checking (i.e. our framework must generate a specification if it is given the expected reference specification). Note how we have almost complete precision for these cases.

We point out that the ChatGPT 4o model has a limited context, that is, if the number of tokens in the learning context is exceeded the context is truncated. For our evaluation steps, our framework has stayed well within this limit. Hence, this sort of context wrapping is not a problem that we are have encountered or are concerned by at the moment.

We also present graphs that depict, at which iteration, functions in the ERCs have been verified. This per-function analysis allows for a more detailed view in the specification generation process. Figures 3, 4, and 5 depict the per-function analysis for ERCs 20, 721, and 1155, respectively. Broadly speaking, for each function, the corresponding (sub-)plot depicts the frequency in which functions have been verified (y-axis) for a given iteration (x-axis). Each colour in the graph represents a different context that has been used to generated the specification for the given ERC. With these graphs, we can see that, as expected, there seems to exist a co-relation between the complexity of a function and its expected specification and how many loop iterations are necessary for it to be verified. Note, for example, how the transferFrom-like functions are more dispersed in these graphs compared to the other functions. Again, this provides additional evidence that our loop-based methodology seems necessary for the generation of specifications for complex functions.

Our results provide encouraging evidence that LLMs can be used to generate formal specification for smart contracts. We have tackled here a class of specifications (i.e. token standards) that are very useful in the domain of smart contracts. Thus, our framework can have already an immediate practical impact in this area. We recognise, however, that this framework and evaluation are only a starting step from which a more thorough investigation should emerge, leading to more robust and general specification-generation frameworks.

5   Related Work

The more general aim of our work is to generate formal models from NL requirements. There are two main approaches to using NLP techniques for writing and processing requirements. One is to define a precise grammar to what is usually denoted as a controlled natural language (CNL), using, for instance, the Grammatical Framework [18]. This obviously facilitates NLP, since the requirements can be parsed and then processed in an unambiguous was, but at the expense of imposing a writing style that users must follow. Some examples are [6,9,16,20].

The other approach is to accept as input free NL requirements without imposing any constraints concerning a well-defined grammar. The advantage is that this is clearly more flexible and is likely to address larger contexts; on the other hand, this tends to be significantly more challenging, exactly due to the inherent ambiguity and variability of natural language. Some examples are [2,19]. We have adopted this more flexible approach, since the intention is to take as input NL specifications of Ethereum standards exactly as they are. Unlike the cited approaches, however, we have based our framework on the use of LLMs.

Unfortunately, there seem to be only a few approaches in this direction. The work in [14] evaluates the precision of a symbolic method and a GPT-based method to generate DbC specifications in the Java Modeling Language (JML) from comments in Java code. We consider here the results related to the GPT-based approach. The GPT model was trained in two stages: a pre-training process on a vast collection of text, followed by a supervised fine-tuning training of the model on a labeled dataset aiming at NLP, but no details of the training are given. Unlike our evaluation, the authors wrote their own (10) NL requirements for their assessment of the models. Although the generated requirements are syntactically and semantically checked, this is not inserted as part of a generation loop. Therefore there is no automatic context improvement from counterexamples, as we do. Out of the 10 requirements, the GPT model could generate only 4 valid JML specifications, although some failed cases were just a matter of missing contextual information.

In [13], the authors investigate the use of ChatGPT to generate loop invariants as annotations for 106 C programs. The authors then used the Frama-C and the CPAchecker verifiers to check the validity (if an invariant holds initially and for each loop iteration), and what they called usefulness (if an invariant helps verifiers in proving tasks). Out of the 106 generated invariants, Chat-GPT managed to generate 75 valid ones; concerning usefulness, Chat-GPT produced 37 useful invariants in comparison with 47 user provided ones. The focus of this work is on reversing engineering formal specifications (particularly loop invariants) from code, whereas we follow a direct engineering approach by generating specifications from NL requirements. Also, despite validity and some form of usefulness checking carried out by the authors, these are not considered in the loop of a(n) (iterative) context improvement as we have done.

6   Conclusion

This paper presented a novel approach to translating natural language into formal specifications for smart contracts based on Large Language Models (LLMs) and, particularly, ChatGPT. The application domain of interest is Ethereum standards, as they are widely used by the blockchain community and have associated textual documentation that we could use, in an unbiased way, to evaluate the DbC-GPT framework we proposed. Our evaluation has shown that it is viable to use ChatGPT for this purpose. Through an iterative process that checks the validity of the generated specifications in the notation of solc-verify, DbC-GPT was able to generate specifications for the 3 standards that were considered, including 8 demonstration context variations. Each variation was formed of a sequence of up-to-three reference specifications for the considered token standards. We were also able to conclude that the more contextual information is provided, the faster a valid specification is generated. More importantly, the iterative process played a crucial role in successfully generating specifications, as, in most cases, valid specifications were obtained only with more than one iteration. As far as we are aware, ours is a first approach to explore this iterative generation of formal specifications from NL descriptions using LLMs.

Looking forward, several opportunities can be explored to enhance and expand the current framework. Our evaluation can be improved to consider many more standards, as well as explore other application domains. The specifications generated by the framework can often be weaker than what is desired; a deeper investigation is necessary and strategies to improve this using appropriate prompt queries is an important topic for future work.

Future work will include using multiple local and non-local LLMs for comparison, such as ChatGPT, Gemini [22], Llama [23], and Claude [21] models, to assess performance differences. We plan to increase the number of experiments and prepare fine-tuning to train models in a standardized manner. Furthermore, we plan to use *a set of implementations* to drive the search for a specification; these implementations can be seen as test cases for the specification being generated, and they should reduce the likelihood of a wrong implementation driving this process to find a wrong specification. Lastly, we will investigate the generation and verification of invariants and preconditions to enhance the robustness and completeness of our framework.

References

1. Solidity language homepage. https://soliditylang.org/. Accessed 14 May 2024
2. Aceituna, D., Do, H., Srinivasan, S.: A systematic approach to transforming system requirements into model checking specifications. In: Companion Proceedings of the 36th International Conference on Software Engineering. p. 165-174. ICSE Companion 2014, Association for Computing Machinery, New York, NY, USA (2014). https://doi.org/10.1145/2591062.2591183, https://doi.org/10.1145/2591062.2591183
3. Antonino, P., Ferreira, J., Sampaio, A., Roscoe, A.W.: Specification is law: Safe creation and upgrade of ethereum smart contracts. In: Schlingloff, B.H., Chai, M. (eds.) Software Engineering and Formal Methods, pp. 227–243. Springer, Cham (2022)
4. Antonino, P., Ferreira, J., Sampaio, A., Roscoe, A., Arruda, F.: A refinement-based approach to safe smart contract deployment and evolution. Software and Systems Modeling, pp. 1–37 (2024)
5. Antonino, P., Roscoe, A.: Solidifier: bounded model checking solidity using lazy contract deployment and precise memory modelling. In: Proceedings of the 36th Annual ACM Symposium on Applied Computing, pp. 1788–1797 (2021)
6. Arruda, F., Barros, F., Sampaio, A.: Automation and consistency analysis of test cases written in natural language: an industrial context. Sci. Comput. Program. **189**, 102377 (2020)
7. Barza, S., Carvalho, G., Iyoda, J., Sampaio, A., Mota, A., Barros, F.: Model checking requirements. In: Ribeiro, L., Lecomte, T. (eds.) SBMF 2016. LNCS, vol. 10090, pp. 217–234. Springer, Cham (2016). https://doi.org/10.1007/978-3-319-49815-7_13
8. Brown, T., Mann, B., Ryder, N., Subbiah, M., Kaplan, J.D., Dhariwal, P., Neelakantan, A., Shyam, P., Sastry, G., Askell, A., et al.: Language models are few-shot learners. Adv. Neural. Inf. Process. Syst. **33**, 1877–1901 (2020)
9. Carvalho, G., Cavalcanti, A., Sampaio, A.: Modelling timed reactive systems from natural-language requirements. Formal Aspects Comput. **28**(5), 725–765 (2016). https://doi.org/10.1007/S00165-016-0387-X
10. Desai, A., et al.: Program synthesis using natural language. In: Proceedings of the 38th International Conference on Software Engineering, ICSE 2016, pp. 345–356. Association for Computing Machinery, New York (2016). https://doi.org/10.1145/2884781.2884786
11. Feng, Z., et al.: CodeBERT: a pre-trained model for programming and natural languages. In: Cohn, T., He, Y., Liu, Y. (eds.) Findings of the Association for Computational Linguistics: EMNLP 2020, pp. 1536–1547. Association for Computational Linguistics, Online, November 2020. https://doi.org/10.18653/v1/2020.findings-emnlp.139. https://aclanthology.org/2020.findings-emnlp.139
12. Hajdu, Á., Jovanović, D.: solc-verify: A modular verifier for solidity smart contracts. In: Verified Software. Theories, Tools, and Experiments: 11th International Conference, VSTTE 2019, New York City, NY, USA, July 13–14, 2019, Revised Selected Papers 11, pp. 161–179. Springer (2020)
13. Janßen, C., Richter, C., Wehrheim, H.: Can chatgpt support software verification? In: Beyer, D., Cavalcanti, A. (eds.) Fundamental Approaches to Software Engineering, pp. 266–279. Springer, Cham (2024)
14. Leong, I.T., Barbosa, R.: Translating natural language requirements to formal specifications: a study on gpt and symbolic nlp. In: 2023 53rd Annual IEEE/IFIP

International Conference on Dependable Systems and Networks Workshops (DSN-W), pp. 259–262 (2023). https://doi.org/10.1109/DSN-W58399.2023.00065

15. Mavridou, A., Laszka, A., Stachtiari, E., Dubey, A.: Verisolid: Correct-by-design smart contracts for ethereum. In: Financial Cryptography and Data Security: 23rd International Conference, FC 2019, Frigate Bay, St. Kitts and Nevis, February 18–22, 2019, Revised Selected Papers 23, pp. 446–465. Springer (2019)

16. Nogueira, S., Sampaio, A., Mota, A.: Test generation from state based use case models. Formal Asp. Comput. **26**(3), 441–490 (2014). https://doi.org/10.1007/s00165-012-0258-z

17. Permenev, A., Dimitrov, D., Tsankov, P., Drachsler-Cohen, D., Vechev, M.: Verx: safety verification of smart contracts. In: 2020 IEEE symposium on security and privacy (SP), pp. 1661–1677. IEEE (2020)

18. Ranta, A.: Grammatical framework. J. Funct. Program. **14**(2), 145-189 (2004). https://doi.org/10.1017/S0956796803004738

19. Santiago Júnior, V.A.D., Vijaykumar, N.L.: Generating model-based test cases from natural language requirements for space application software. Softw. Quality J. **20**(1), 77–143 (2012).https://doi.org/10.1007/s11219-011-9155-6

20. Selway, M., Grossmann, G., Mayer, W., Stumptner, M.: Formalising natural language specifications using a cognitive linguistic/configuration based approach. Inf. Syst. **54**, 191–208 (2015). https://doi.org/10.1016/j.is.2015.04.003, https://www.sciencedirect.com/science/article/pii/S0306437915000630

21. team, C.: Claude: A next-generation ai assistant by anthropic. https://www.anthropic.com/claude. Accessed 19 July 2024

22. Team, G., et al.: Gemini: a family of highly capable multimodal models. arXiv preprint arXiv:2312.11805 (2023)

23. Touvron, H., et al.: Llama 2: Open foundation and fine-tuned chat models. arXiv preprint arXiv:2307.09288 (2023)

24. Vaswani, A., Shazeer, N., Parmar, N., Uszkoreit, J., Jones, L., Gomez, A.N., Kaiser, Ł., Polosukhin, I.: Attention is all you need. Advances in neural information processing systems **30** (2017)

25. Wang, C., Pastore, F., Goknil, A., Briand, L.C., Iqbal, Z.: Umtg: a toolset to automatically generate system test cases from use case specifications. In: Proceedings of the 2015 10th Joint Meeting on Foundations of Software Engineering, pp. 942–945. ESEC/FSE 2015, ACM, New York (2015). https://doi.org/10.1145/2786805.2803187

26. Wang, Y., Lahiri, S.K., Chen, S., Pan, R., Dillig, I., Born, C., Naseer, I.: Formal specification and verification of smart contracts for azure blockchain. arXiv preprint arXiv:1812.08829 (2018)

How Do Asynchronous Communication Models Impact the Composability of Information Flow Security?

Lena Gerlach[1] and Christopher Gerking[2]([⊠]) [iD]

[1] Technical University of Berlin, Berlin, Germany
`lena.gerlach@campus.tu-berlin.de`
[2] Karlsruhe Institute of Technology (KIT), Karlsruhe, Germany
`christopher.gerking@kit.edu`

Abstract. Information flow security is not guaranteed to be preserved when a system is being composed of secure components. Whereas asynchronous communication is known to mitigate this problem in principle, the concrete impact of specific communication models on composability has not been investigated so far. We address this problem by formalizing seven asynchronous communication models from the literature in the UPPAAL environment. On this basis, we capture a negative example, where information flow security is not composable, and contrast it with the formalized communication models to investigate their impact on composability. Our investigation shows that six out of seven asynchronous communication models ensure composability of information flow security.

Keywords: Information Flow Security · Asynchronous Communication

1 Introduction

Information security is a quality property that benefits greatly from formal methods, as their application enables even subtle information leaks to be detected. The theory of *information flow security* [17] provides such formal methods, with *noninterference* being the most prominent formal security property [15]. Whereas the verification of properties like noninterference is a long-established research topic, their *composability* remains challenging: opposed to properties like safety or liveness, information flow security is a *hyperproperty* [9] that is not guaranteed to be preserved on composition. Thus, a composite system may leak information even though it is composed of secure components only [18]. This challenge gains practical relevance in areas such as *component software* [23], where the preservation of quality properties on composition of components plays a key role.

In principle, it is known that asynchronous message passing between components enables information flow security to be preserved on composition [24].

L. Gerlach—The research was conducted when the author was with KIT.

However, communication models for message passing are diverse [8,14], ranging from restrictive models (resembling synchronous communication) to more liberal, fully-fledged asynchronous models. In this paper, we address the problem that the impact of these different models on the composability of information flow security was not investigated so far. Despite the well-known fact that asynchronous communication has a positive impact on the preservation of information flow security in principle, it is still unclear which concrete characteristics of message passing facilitate the composability, or give rise to information leaks.

Related work deals with the verification of hyperproperties like information flow security [16]. Here, the focus is on the security of individual components, not on preserving security when a system is being composed of secure components. Another work verifies compatibility criteria of asynchronously communicating components within a composition [6]. Thereby, the approach disregards information flow security or other hyperproperties, as well as their preservation. Finally, another work actually deals with preserving properties under asynchronous communication [5]. However, unlike our case, the preservation refers to changing a composed system from one communication model to another. Since the approach is focused on the preservation of properties like safety, it supports no final conclusions about hyperproperties like information flow security.

In this paper, we report on the lead author's thesis [12] which investigated how asynchronous communication models from literature impact the composability of information flow security. To this end, we formalize a set of seven communication models considered by Chevrou et al. [8], making them amenable to formal verification by means of the UPPAAL environment. As our information flow property under investigation, we restrict ourselves to *generalized noninterference* (GNI) [20] as it is known to be composable under asynchronous communication in principle [24]. On this basis, we construct the working hypothesis that GNI is in fact composable under the majority (but not all) of the investigated communication models. To validate our hypothesis, we use UPPAAL to formalize a negative example from literature [20], in which GNI fails to be composable under purely synchronous communication. To uncover the positive impact of asynchronous communication, we contrast the example with each of the seven communication models, using UPPAAL to investigate the composability of GNI.

In summary, this paper makes the following contributions:

- We provide reusable formalizations of seven asynchronous communication models in the UPPAAL environment.
- We represent a negative example in UPPAAL to investigate the impact of communication models on the composability of information flow security.
- We use formal verification with UPPAAL to show that six out of seven communication models ensure composability.

Paper Organization: We provide fundamentals on asynchronous communication models, information flow security, and UPPAAL in Sect. 2. In Sect. 3, we describe the formalization of communication models in UPPAAL. In the subsequent Sect. 4, we outline related work on the composability of information flow security and construct our working hypothesis. We validate our hypothesis by investigating the composability in Sect. 5, before concluding in Sect. 6.

2 Fundamentals

This chapter delves into the key fundamentals of this work, which are essential for understanding and contextualizing the subsequent content. In Sect. 2.1, we discuss asynchronous communication models underlying our work. This is followed by a definition of formal security in terms of information flow in Sect. 2.2. Finally, Sect. 2.3 introduces the UPPAAL verification tool.

2.1 Asynchronous Communication Models

We investigate the nuances between synchronous and asynchronous communication and their impact on the composability of information flow security. In asynchronous communication, the sending and receiving of messages are decoupled and do not occur simultaneously. In contrast, synchronous communication is characterized by atomic events, i.e., sending and receiving must occur simultaneously. Additionally, synchronous communication is blocking: to send a message, the sender requires a receiver that is actually ready-to-receive.

Intermediate models between synchronous and asynchronous communication enable non-atomic sending and receiving of messages. Chevrou et al. [7] distinguish between seven diverse forms of asynchronous communication. Four of them are different first-in-first-out (FIFO) variants using message buffers managed in a FIFO manner. The FIFO variants differ in the coordination of the communicating components, which we refer to as *peers*. In addition to FIFO models, the communication can be *realizable with synchronous communication* (RSC), Causal, or fully asynchronous (Async). We summarize the conditions for the individual asynchronous models as follows:

- **RSC:** A new message can only be sent after the previously sent message has been received. The receipt follows the sending immediately consecutive. However, unlike synchronous communication, RSC does not enforce atomic sending and receiving, as they take place one after another.
- **FIFO N-N:** Coordinates all senders and receivers. All messages are globally sorted in a single message buffer and delivered in the order they were sent.
- **FIFO N-1:** Coordinates a single receiver with all senders. Each receiver has a local input buffer in which all messages sent from different senders to this receiver are stored in the order they were sent.
- **FIFO 1-N:** Coordinates a single sender with all receivers. Each sender has a local output buffer from which all messages addressed to different receivers are delivered in the order they were sent.
- **Causal:** There is a global message buffer for all senders and receivers, managed according to causal dependencies.
- **FIFO 1-1:** Coordinates a single sender with a single receiver. Each sender-receiver pair has a message buffer, whereas messages exchanged between sender and receiver are received in the order they were sent.
- **Async:** There is no order in message delivery; messages can be skipped or delayed arbitrarily. There is a single global message buffer for all senders and receivers, without any management constraints.

According to Chevrou et al. [7], a communication model K_1 is contained within another communication model K_2 if all valid executions of K_1 are also valid executions of K_2. This forms a hierarchy that demonstrates the strictness of the conditions imposed by each communication model. For instance, Async has the weakest conditions as it does not impose an order in which messages are to be delivered. The remaining communication models impose stricter conditions, making them subsets of Async. Whereas Chevrou et al. [7] characterize FIFO 1-N and FIFO N-1 as incomparable, Di Giusto et al. [14] present an alternative hierarchy in which they represent FIFO 1-N as a subclass of FIFO N-1.

2.2   Information Flow Security

In this section, we introduce secure information flow and the information flow property that we use to analyze the impact of various communication models on composability. Following Clarkson and Schneider [9], information flow security imposes restrictions on the information that users can learn by interacting with a system. They characterize information flow properties as *hyperproperties* of a system, which encompass a set of trace properties. A trace property describes a set of execution traces that all satisfy a specific property. Trace properties consist of *safety* and *liveness* properties. A *safety* property ensures that nothing "bad" happens during execution, while a *liveness* property ensures that something "good" happens during execution. Since hyperproperties represent a collection of trace properties, they cannot be satisfied by a single trace alone. Instead, multiple traces are needed, each satisfying a trace property within the hyperproperty.

The best known information flow property is *noninterference* [15]. It specifies that the public (or "low") outputs of a system must not depend on confidential (or "high") inputs. Whereas noninterference is restricted to deterministic state machines, generalized noninterference (GNI) is a nondeterministic extension. For *Prop* as the set of all trace properties, GNI is formally defined as follows [9]:

$$GNI \;\hat{=}\; \{T \in Prop \mid (\forall t_1, t_2 \in T : (\exists t_3 \in T : ev_{Hin}(t_3) = ev_{Hin}(t_1) \\ \land \; ev_L(t_3) = ev_L(t_2)))\} \tag{1}$$

Defining GNI as a set in Eq. (1) demands that for all traces t_1 and t_2, there must exist a third trace t_3 that has the same confidential input events (Hin) as t_1 and the same public events (L, including inputs and outputs) as t_2. When observing the public events of t_3, it is therefore not deducible whether the confidential inputs of t_1 were actually received (on t_3) or not (on t_2). We refer to a system with its set of traces T being included in the GNI set as *GNI-secure*.

2.3   UPPAAL

In this work, we use the model checking tool UPPAAL [3] to investigate the impact of communication models formally. UPPAAL was developed to verify real-time systems modeled as timed automata, which are frequently used as a formalism to express security [2]. Accordingly, we assume that the message passing behavior of components is formalized using UPPAAL timed automata.

Timed Automata. Extend finite automata by incorporating a finite set of clocks to track the passage of time. A timed automaton is defined as a tuple $\langle \Sigma, S, S_0, C, E \rangle$, where:

- Σ is an alphabet of symbols
- S is a finite set of locations
- $S_0 \subseteq S$ is a set of initial locations
- C is a set of clocks
- $E \subseteq S \times S \times [\Sigma \cup \{\epsilon\}] \times 2^C \times \Phi(C)$ is a set of edges, ϵ the internal action [1]

$\Phi(C)$ is a set of formulas con-structed from atomic formulas $x \le c$ or $c \le x$, where $x \in C$ and $c \in N$ with N being a set of time constants. An edge $\langle s, s', \sigma, \lambda, \delta \rangle \in E$ as illustrated in Fig. 1 represents a state transition

Fig. 1. Syntax of timed automata

from s to s' triggered by the symbol σ. The set λ specifies the clocks that must be reset during the transition, meaning the clocks start counting from zero again. δ is a formula from $\Phi(C)$ that serves as a guard, representing the condition that must be satisfied to execute the state transition. When editing an edge in UPPAAL, the following labels can be used in addition to the aforementioned guards:

Sync: Corresponds to σ from the aforementioned tuple and is used as a synchro-nization channel to compose multiple timed automata into a network based on the *handshaking* approach [4].
Update: Extends λ from the tuple, allowing the specification of expressions with side effects. In particular, updates may invoke predefined functions.
Select: Used to define variables that can take a nondeterministic value within their type range and can only be accessed on the corresponding edge. These variables can be used by expressions defined as guard or update.

Computation Tree Logic (CTL). Extends propositional logic with a tempo-ral aspect, allowing the truth value of statements to reference the future or past. This enables the specification of queries considering changes in truth values over time. As a branching time logic, the modeling of time in CTL corresponds to a tree structure with branches. UPPAAL employs a simplified form of CTL [3], consisting of path formulas and state formulas. State formulas describe individ-ual states, while path formulas quantify over paths. For state formulas **p** and **q**, UPPAAL enables the specification of *reachability*, *safety*, and *liveness* properties.

Reachability properties:
- **E<> p**: Satisfied if there exists a path where **p** is true in some state
Safety properties:
- **A[] p**: Satisfied if all paths include only states in which **p** is true
- **E[] p**: Satisfied if there exists a path where **p** is true in all states
Liveness properties:
- **A<> p**: Satisfied if all paths include some state in which **p** is true
- **p --> q**: Satisfied if, whenever a state exists in which **p** is true, there is a subsequent state on all paths from this state in which **q** is true

3   Formalization of Asynchronous Communication

In order to investigate the impact of different communication models on the composability of GNI, it is necessary to formalize these models. Our formalization is based on a concept presented in our preliminary work [10], which was designed to enable asynchronous message passing between timed automata in UPPAAL. In general, the concept decouples message sending from receiving by buffering sent messages until they are consumed by the intended receiver. The communication modeled in the original concept corresponds to a FIFO N-1 communication as described in Sect. 2.1, which is a restriction that we loosen within this paper.

As part of the underlying concept, components act as communicating peers, which we assume to be uniquely identifiable by means of integer values. Similarly, messages are encoded by integer values as well, assuming that distinct messages of the same type are indistinguishable. On this basis, we encapsulate the asynchronous message passing by means of uniform interfaces that consist of four core operations implemented by each communication model:

Requesting a message of a specific type to be sent to a dedicated receiving peer. The corresponding operation is `request(Peer sender, Message m, Peer receiver) : bool`, which is invoked by the `sender` as a guard on one of its edges. The invocation returns true if message `m` can be passed to the communication model, as it complies with the model's conditions to be sent.

Sending a message of a specific type from the sending peer to a dedicated receiving peer. The corresponding operation is `send(Peer sender, Message m, Peer receiver)`, which is invoked by the `sender` as an update on one of its edges. The invocation passes the message `m` to the communication model, so that it will be eventually received by the peer `receiver`.

Receiving a message of a specific type from a dedicated sending peer. The corresponding operation is `receive(Peer receiver, Message m, Peer sender) : bool`, which is invoked by the `receiver` as a guard on one of its edges. The invocation returns true if message `m` was passed to the communication model by the peer `sender`, and complies with the model's conditions to be received.

Consuming a message of a specific type upon receipt. The corresponding operation is `consume(Peer receiver, Message m, Peer sender)`, which is invoked by the `receiver` as an update on one of its edges. The invocation removes message `m` from the communication model.

In Fig. 2, we illustrate the general concept using two components as peers P_1 and P_2. As depicted, we assume that the component behavior is formalized using timed automata (cf. Sect. 2.3). According to the behavior shown in Fig. 2, messages are sent by P_1 and received by P_2. This is not a general restriction, since we assume that components act as both sender and receiver. In the particular case shown, P_1 requires an interface for requesting and sending, whereas P_2 requires an interface for receiving and consuming. Both interfaces are provided by a communication model that implements the associated operations in one of

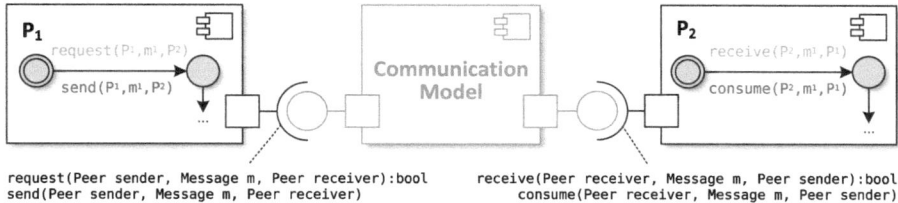

Fig. 2. Composition of two peers P_1 and P_2 under a communication model

various ways. Theoretically, different pairs of components can also use different models, while our implementation is currently limited to a single model at each time.

A commonality shared by all communication models is the use of buffers to decouple message sending from receiving. Accordingly, basic functionalities of the **request**, **send**, **receive**, and **consume** operations remain consistent across all models: sending a message is preceded by a request to add it to the respective buffer, while receiving a message means checking for its presence in the corresponding buffer before subsequently consuming it. In UPPAAL, buffers are formalized as arrays, with buffer sizes and the number of buffers tailored to specific models. According to the operation parameters depicted in Fig. 2, the message (denoted as **m**) and the communicating peers must always be specified when invoking any of the operations. In general, sender and receiver parameters (each of type **Peer**) identify the respective buffers used by the communication model to store messages between request/sending and receipt/consumption.

Below, we outline our formalization of the communication models introduced in Sect. 2.1. The formalizations differ only in the implementation of the four operations **request**, **send**, **receive**, and **consume**. By encapsulating the implementation details, we ensure that changing the communication model does not require modifications to the components invoking the operations, thereby upholding the principle of *information hiding* [21]. In the following, we summarize the functionalities associated with each operation across different communication models, elaborating on the RSC model as it makes full use of all operations. In UPPAAL, we implement the operations in terms of functions as part of the global declarations [3]. Thereby, we ensure that they are globally accessible from all automata representing the behavior of components. For insights into the implementation details, we refer the reader to our reproduction package [13].

RSC. We implement the RSC model using a single global buffer that is shared between all peers, with a capacity limited to one message. This model operates in a blocking manner, allowing a message to be sent only when the buffer is available. To manage the availability status, it is sufficient to increment and decrement the buffer's **tail** field, which is used to indicate whether the buffer is free or occupied. Upon a request to send, the implementation of the **request** function checks whether the **tail** value is less then the capacity of 1 (cf. listing

1.1). On that condition, the **send** function places a message into the buffer and marks it as occupied by incrementing the **tail**.

Listing 1.1. Formalization of **request** and **send** under the RSC model

```
bool request(Peer sender,        void send(Peer sender,
Message m, Peer receiver) {      Message m, Peer receiver) {
    return (m != null and           buffer.messages[0] = m;
    buffer.tail < 1);               buffer.tail++;
}                                }
```

According to the fixed capacity of 1, only the message at position 0 can be received by a peer invoking the **receive** function. In that case, the **consume** function removes the received message from the buffer and overrides its position 0 with a null value (cf. Listing 1.2), thereby updating the buffer's status to free. Given that peers communicate through one shared buffer, the individual identifiers of the sender and receiver are not taken into account by any of the functions illustrated in listing 1.1 and 1.2. This is because it is inherently clear which buffer a message is sent to or received from.

Listing 1.2. Formalization of **receive** and **consume** under the RSC model

```
bool receive(Peer receiver,      void consume(Peer receiver,
Message m, Peer sender) {        Message m, Peer sender) {
    return (m != null and           buffer.messages[0] = null;
    buffer.messages[0] == m);       buffer.tail--;
}                                }
```

FIFO N-N. The implementation of the FIFO N-N model employs a single central buffer as well, which is managed according to a FIFO strategy. Unlike RSC, this buffer may store more than one message (nevertheless being limited to a fixed maximum capacity). The FIFO management ensures that messages are received, stored, and consumed in the sequence they were sent. Since the FIFO N-N communication is non-blocking, the **request** operation will constantly return true (even if the buffer is full). Whereas this implementation varies significantly from RSC, it represents a commonality that FIFO N-N shares with all further models described below.

FIFO N-1. In the FIFO N-1 model, the number of buffers matches the number of peers. Thereby, we generally assume that all peers actually participate as receiver of messages. Accordingly, each peer is associated with its own local receive buffer managed under the FIFO principle. Therefore, the **send** function utilizes the **receiver** parameter to identify the designated buffer into which messages must be inserted. Accordingly, upon receipt and consumption of messages, peers access only their own local receive buffers. To that end, the **receiver** parameter is also crucial for the `receive` and `consume` functions to identify the correct buffer, whereas the `sender` parameter is not taken into account.

Fifo 1-N. In the Fifo 1-N model, the number of message buffers aligns with the number of peers once more. This decision is based on the general assumption that all peers act as senders of messages. Accordingly, unlike the Fifo N-1 model, buffers act as output buffers on the sender side. Therefore, sending a message involves adding it to the buffer that corresponds to the sender. The `send` function thus requires the `sender` parameter to indicate the designated buffer. Similarly, the `receive` and `consume` functions utilize the `sender` parameter to retrieve messages from the senders' buffers, whereas the `receiver` parameter is not utilized by any of these functions.

Causal. Causally ordered communication ensures that if message m_1 is causally sent before another message m_2 with the same receiver, then m_2 cannot be consumed prior to m_1. *Vector clocks*, as described by Chevrou et al. [8], provide a means to achieve this causal ordering. For each peer P_i, the communication model handles a vector clock denoted as $vcOf(P_i)$, where the number of vector components aligns with the number of peers. For each peer P, the vector $vcOf(P)(P)$ corresponds to its own vector component within its respective vector clock. Specifically, for peers P_1 and P_2, the vector $vcOf(P_1)(P_2)$ denotes the count of send events by P_2 that are in the causal past of P_1. This indicates the number of messages sent by P_2 and known to P_1.

A message m_1 is considered in the causal past of m_2 if every vector component of m_1 is less than or equal to the corresponding vector component of m_2, with at least one component being strictly smaller. In addition to vector clocks, similar to Fifo N-1, each peer maintains a receive buffer. When a message is sent using the `send` function, the sender's own component of the vector clock is incremented. The updated vector clock is then sent along with the message to the receiver's buffer. Upon adding a message to the receive buffer, it is sorted into the correct position within the buffer. The correct position means that all messages at positions with a smaller index possess smaller vector values, while those at positions with a larger index possess larger vector values. The causal ordering of message receipts is ensured by receiving messages at index 0, adhering to the Fifo principle. During message consumption, each component of the receiver's vector clock is updated with the maximum of its current value and the value of the component of the received vector clock.

Fifo 1-1. In the Fifo 1-1 model, each possible pair of peers is associated with a dedicated buffer. For clarity, we omit buffers shared by a peer with itself, as peers are not assumed to communicate internally. Peers send messages to the specific buffers shared with their designated receivers. The `send` function therefore requires both `sender` and `receiver` parameters to identify the correct buffer for message transmission. Similarly, messages will be received and consumed only from those buffers shared between sender and receiver. Consequently, to access the appropriate buffer, the `receive` and `consume` functions also depend on the `sender` and `receiver` parameters.

Async. In fully asynchronous communication, a global message buffer operates without any imposed restrictions. Unlike the FIFO models described above, messages can be accessed from any position within the buffer, not just the first. During message receipt, the entire buffer is checked if necessary, rather than position 0 only. Nevertheless, to account for the fact that functions in UPPAAL require a deterministic behavior, our implementation of the `consume` function will always extract the first occurrence of the specified message type from the buffer. Thereby, we rely on our assumption that messages are represented solely by integers. For messages with the same type, the order of occurrence within the buffer is therefore irrelevant. Similar to RSC and FIFO N-N, fully asynchronous communication shares a single central buffer between all peers. Therefore, the `sender` and `receiver` parameters are not utilized in any of the functions.

4   Related Work and Formation of Hypothesis

We survey related work to hypothesize the impact of asynchronous communication models on the composability of GNI. The hypothesis formed in Sect. 4.2 is derived from the literature overview given in the preceding Sect. 4.1.

4.1   Literature Overview

McCullough [20] was the first to demonstrate that composing two GNI-secure components can result in a system that no longer satisfies GNI. The type of composition in his example is referred to as *feedback*, as it enables both components to act as sender and receiver at the same time. Outputs of one component act as immediate inputs to the other, thereby suggesting synchronous communication. Nevertheless, since McCullough does not explicitly consider the communication model, it remains unclear how different models may impact the preservation of GNI. In a later work [19], McCullough highlights the impact of blocking message buffers on GNI composability. When a buffer is full, attempting to send another message to it can provide information to the sender. Subsequently, when the sending is successful, the sender may conclude that another component has received a message, thereby potentially disclosing confidential information.

Zakinthinos and Lee [24] build upon McCullough's findings by demonstrating the preservation of GNI in compositions without feedback. Additionally, they propose a condition to preserve GNI even in compositions with feedback, which is the delay of confidential messages. In particular, it must be possible to delay the receipt of a confidential message until after a subsequent public message is sent or received. Such a delay enables public messages to be sent regardless of whether specific confidential messages were previously received or not. Accordingly, no information can be deduced from the public messages as to whether or not confidential messages were received, thus ensuring GNI. On this basis, the authors identify asynchronous communication as an approach that facilitates this delay. Accordingly, they distinguish between synchronous and asynchronous communication, however, without going into the nuances between them.

Sabelfeld [22] investigates the security implications of information flow in multi-threaded systems, focusing on process synchronization. He introduces a type-based composition analysis to verify program security, demonstrating that security requirements (e.g., the composability of GNI) impose constraints on synchronized processes. In contrast to our work, Sabelfeld investigates secure information flow on the basis of program code, rather than abstract behavioral descriptions such as automata. Moreover, since Sabelfeld's analysis does not include asynchronous communication, he does not look closely at how varying degrees of synchronicity impact secure information flow.

Lamport and Schneider [16] use the Temporal Logic of Actions (TLA) to verify hyperproperties, GNI among them. Whereas the authors thereby offer formal guarantees for information flow security, they do not address a compositional verification that enables global guarantees to be composed from local guarantees of individual components. In contrast, Chevrou et al. [6] employ TLA as well, addressing the verification of asynchronously communicating systems. Whereas the authors consider the full range of communication models that is underlying our work, they focus solely on the compatibility of components within a composition. Accordingly, they verify trace properties like termination or non-blockage, rather than hyperproperties like information flow security. Bunte et al. [5] address the preservation of properties when changing a composition from one communication model to another. Thereby, they do not focus on creating new compositions, which is the subject of our work. By guaranteeing safety properties and identifying conditions for reachability, deadlock freedom, or confluence, the authors do not deal with hyperproperties like information flow security.

4.2   Research Hypothesis

On the basis of the literature discussed in Sect. 4.1, we form a research hypothesis about the impact of asynchronous communication models from Sect. 3 on the composability of GNI. In particular, we make reference to the fundamental work by Zakinthinos and Lee [24] as well as McCullough [19]. In all asynchronous communication models discussed in Sect. 3, a message delay as described by Zakinthinos and Lee [24] can be granted, with the exception of RSC. Assuming that a confidential message has been sent from one component to another, the RSC model stipulates that the subsequent event is the receipt of the exact same message. Even though the receipt can be delayed arbitrarily unlike in synchronous communication, RSC prevents additional messages from being sent during this delay. In contrast, all other models described in Sect. 3 enable the message receipt to be delayed without preventing other public messages from being sent in the meantime.

The above observation is in line with McCullough [19], who points out the blocking of message buffers as an influencing factor for the composability of GNI. Among the communication models formalized in Sect. 3, RSC is the only one that potentially blocks a component when requesting a message to be sent (by returning `false` from the `request` function whenever the buffer is already occupied). Conversely, all other models operate on a default implementation that returns true regardless of whether the buffer is occupied or not. Given that the buffer is blocking under the RSC model and that RSC does not enable confidential messages to be delayed, we form the following research hypothesis.

> **Hypothesis**: When composing two GNI-secure systems, the asynchronous communication models FIFO N-N, FIFO N-1, FIFO 1-N, Causal, FIFO 1-1, and Async preserve GNI. In contrast, the RSC communication model does not ensure composability of GNI.

5   Investigation of Composability

In the following, we intend to corroborate our hypothesis formed in Sect. 4.2, thereby responding to the research question addressed in the title of this paper. To identify metrics for the composability of GNI, we select an example from literature in which two systems fail to preserve GNI on composition with synchronous communication (cf. Sect. 5.1). We formalize this negative example in UPPAAL while replacing synchronous with asynchronous communication (cf. Sect. 5.2). Next, we evaluate the preservation of GNI for each communication model (cf. Sect. 5.3), before discussing obtained results (cf. Sect. 5.4) and threats to validity (cf. Sect. 5.5).

5.1   Negative Example with Synchronous Communication

For a negative example, we refer to McCullough [20] who presents a composition of two GNI-secure systems, A and B, resulting in an overall system that is no longer GNI-secure. System A comprises two classes of messages: confidential messages (which we denote by secret) and public messages. To indicate that confidential inputs of type secret were processed internally, the system answers with a corresponding secret output. Furthermore, it is possible to remove the confidential messages received by A before they are processed internally. To this end, a public cancel input is sent to A. To confirm the receipt, the system answers with a public cancel output. But if no secret input was received prior to the cancel input, or if the confidential messages have already been processed internally, system A sends another public output of type *nothing to cancel* (which we abbreviate as nothing). System B is identical to A, except that it does not confirm the cancel input with another cancel output. Thereby, it prevents the cancel messages from being sent back and forth between A and B infinitely.

In the following, we illustrate that A and B are GNI-secure according to Eq. (1): if a secret input is followed by a cancel input, either the confidential messages are removed and nothing is not sent (t_1), or they are processed and nothing *is* sent (t_3). Instead, if no secret input is received (t_2), the cancel input is followed by a nothing output (like t_3). Hence, A and B send nothing (on t_2 and t_3) regardless of whether secret was received (on t_1) or not (on t_2).

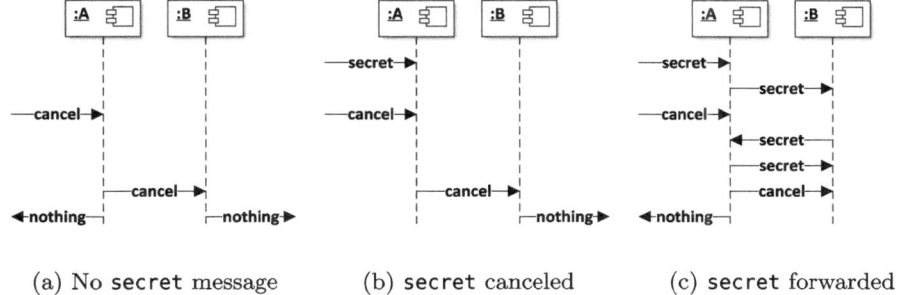

(a) No secret message　　　(b) secret canceled　　　(c) secret forwarded

Fig. 3. Synchronous communication of components A and B on composition [20]

Thus, it is impossible to deduce from the public output whether confidential inputs were received, thereby illustrating that both systems satisfy GNI.

McCullough composes the two systems as shown in Fig. 3: outputs from A become inputs to B and vice versa (except nothing, which is output to the environment). Unlike B, system A receives additional inputs from the environment. On this basis, McCullough presents three execution traces to demonstrate that the composite system of A and B is not GNI-secure under synchronous communication. In Fig. 3a, the composite system receives a cancel input only, resulting in two nothing outputs. When adding a secret input prior to the cancel input, multiple traces are possible due to nondeterministic behavior: in Fig. 3b, secret is not processed by A and thus not forwarded to B. If so, A does not output nothing while B does. Alternatively, in Fig. 3c, secret *is* processed and forwarded to B. If so, A outputs nothing but B does not. Using Fig. 3c, McCullough illustrates that for a trace with the same secret input as in Fig. 3b, it is impossible to include the same number of two nothing outputs as in Fig. 3a [20]. When the composite system sends a single nothing output, it is therefore deducible that it has received a secret input, which represents a violation of GNI.

5.2　Formalization in UPPAAL

To formalize the description of A and B, we inspect the traces in Fig. 3 and identify possible sequences of send and receive events. Starting from an initial location, we convert these sequences into UPPAAL timed automata with edges defining the communication events for each individual location.

Figure 4 shows the resulting automaton for A, with horizontal handling of secret messages and vertical handling of cancel or nothing messages. To handle messages as described in Sect. 3, the request, send, receive, and consume functions are invoked as updates or guards in UPPAAL (cf. Sect. 2.3). Since messages like secret can be received from different senders (cf. Fig. 3), some edges use UPPAAL's select feature (cf. Sect. 2.3) to accept an arbitrary sender nondeterministically. In particular, a cancel output is sent after a cancel input

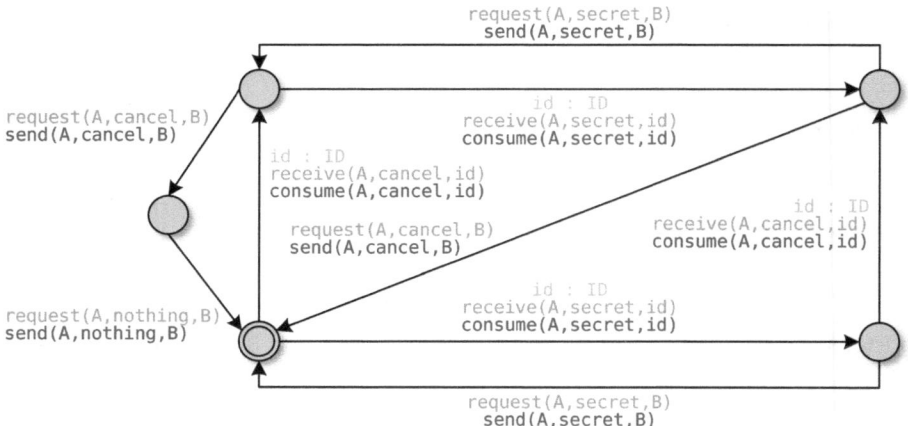

Fig. 4. UPPAAL timed automaton of the system **A**

was received. The automaton for **B** is identical, except that it does not send `cancel` back to **A**. To complement our automata with synchronous communication, we replace the aforementioned functions with synchronization channels (cf. Sect. 2.3).

5.3   Evaluation

This section presents our UPPAAL queries to evaluate the composability of GNI. The queries Q1–Q7 in Table 1 are reachability or safety properties (cf. Sect. 2.3) based on two global variables: the boolean `isSecret` is set to true when a `secret` message is received, while the integer variable `nothingCount` tracks the number of `nothing` messages sent. Due to space restrictions, we refer to our reproduction package for details on the integration of these variables into our automata [11].

On this basis, Q1 checks for a trace on which `secret` is received, yet `nothing` is sent twice. As discussed in Sect. 5.1, this is a prerequisite for the composite system to be GNI-secure. If unsatisfied, Q1 therefore serves as a counterexample to disprove the preservation of GNI. In contrast, satisfying Q1 can only present arguments in favor of the preservation (without formally proving it). To ensure that **A** and **B** are themselves GNI-secure, we introduce two local boolean variables `_isSecret` and `_isNothing`, set to true when **A** and **B** receive `secret` or send `nothing`. Thereby, Q2–Q3 check **A** for traces on which `nothing` is sent whether or not `secret` is received. Q4 checks whether `nothing` is possibly never sent, otherwise indicating that **A** was constructed incorrectly. If Q2–Q4 are satisfied, we assume **A** to be GNI-secure. Similarly, we check **B** for GNI using Q5–Q7.

5.4 Results

Table 1 summarizes the results obtained from our evaluation scheme described in Sect. 5.3. When checking the negative example against our queries, we find that all asynchronous communication models from Sect. 3 satisfy Q2–Q7. The same applies to a synchronous communication model (Sync), which we include as a plausibility check. In contrast, Q1 is neither satisfied by synchronous communication, nor by asynchronous communication using the RSC model.

Table 1. Queries and verification results under different communication models

Queries		RSC & Sync	FIFO 1-1 − Async
Q1	E<> isSecret == **true** and nothingCount == 2	✗	✓
Q2	E<> A._isSecret == **true** and A._isNothing == **true**	✓	✓
Q3	E<> A._isSecret == **false** and A._isNothing == **true**	✓	✓
Q4	E[] A._isNothing == **false**	✓	✓
Q5	E<> B._isSecret == **true** and B._isNothing == **true**	✓	✓
Q6	E<> B._isSecret == **false** and B._isNothing == **true**	✓	✓
Q7	E[] B._isNothing == **false**	✓	✓
Generalized noninterference (GNI)		✗	✓

Since all communication models satisfy Q2–Q7, we assume systems A and B to be GNI-secure. The violation of Q1 aligns with the findings from literature [19] that synchronous communication does not preserve GNI. The fact that Q1 is satisfied for fully asynchronous communication is consistent with the literature as well [24]. In contrast to all other asynchronous models, RSC does not satisfy Q1 and thus violates the composability of GNI, which corroborates our hypothesis from Sect. 4.2. Since both RSC and Sync are blocking, we thereby confirm non-blocking communication as a condition for preserving GNI on composition. Furthermore, we also confirm the condition that it must be possible to delay the receipt of a confidential message until another public message is sent or received.

For the asynchronous models satisfying Q1, Fig. 5 shows an exemplary trace. It becomes apparent that the composability of GNI requires B to delay the receipt of the secret input long enough for the nothing output to be sent first. Without this delay, B would receive the secret message first, thereby preventing its nothing output.

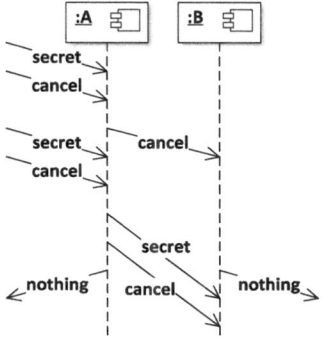

Fig. 5. Trace in asynchronous models satisfying query Q1

5.5 Validity

A threat to the construct validity is that we rely on an example case (cf. Sect. 5.1). While a single case can show that a communication model violates the composability of GNI, it does not provide formal evidence for ensuring composability in all cases. In addition, while query Q1 can disprove the composability, it does not conclusively prove that a communication model preserves GNI, not even for an individual example case. This limitation is caused by the fact that GNI is a hyperproperty (cf. Sect. 2.2), whereas UPPAAL is restricted to the verification of trace properties like reachability (cf. Sect. 2.3).

The internal validity of our findings is threatened by the fact that McCullough's example case was not initially given in a formalized form [20]. Errors may have occurred during its formalization (cf. Sect. 5.2), potentially compromising the requirement that A and B are GNI-secure. We counteracted this threat by verifying the queries Q2–Q7. Nevertheless, a violation of GNI might still be caused by the composed systems, rather than by the communication model.

Regarding external validity, the generalizability of the results is limited. Only GNI was considered as an information flow property, so the findings cannot be transferred to other relevant properties that would be interesting to investigate as well. In terms of reliability, we published our formalization of the communication models as part of a reproduction package [11]. The UPPAAL implementation of the negative example and the associated queries are detailed in Sect. 5.2. Thus, the formalization of communication models, the representation of the negative example in UPPAAL, as well as the evaluation scheme are reproducible.

6 Conclusion and Future Work

In this paper, we investigated the impact of asynchronous communication models on the composability of information flow security. To that end, we formalized various communication models in the UPPAAL environment, introducing a uniform interface to encapsulate their functional varieties. On this basis, we referred to a negative example from literature, in which GNI fails to be composable, and substituted synchronous with asynchronous communication. We applied UPPAAL formal verification to confirm our research hypothesis that all asynchronous communication models, except the RSC model, ensure composability of GNI.

Even though our results basically confirm the well-known difference between synchronous and asynchronous communication in the field of information flow security [24], they are nevertheless meaningful to the diverse field of asynchronous communication models [7]. First, our formalization of communication models in UPPAAL can be reused by theorists during formal verification. Second, our results are novel for practitioners putting a communication model in place while reflecting upon information flow security and how it might be impacted. Finally, our results are of relevance to providers of existing or developers of new communication models willing to inform clients about the impact on security.

In future work, we intend to strengthen our results by proving hyperproperties like information flow security formally under different asynchronous communication models. Furthermore, we would like to consider the real-time behavior of communication models, which is why our work is based on timed automata in the UPPAAL environment. Even though we did not make use of its real-time capabilities so far, we intend to take into account the transmission time of messages and the induced time delays explicitly in our future research.

Acknowledgments. This work was funded by the Topic Engineering Secure Systems of the Helmholtz Association (HGF) and supported by KASTEL Security Research Labs, Karlsruhe.

References

1. Alur, R., Dill, D.: Automata for modeling real-time systems. In: Paterson, M.S. (ed.) ICALP 1990. LNCS, vol. 443, pp. 322–335. Springer, Heidelberg (1990). https://doi.org/10.1007/BFb0032042
2. Arcile, J., André, É.: Timed automata as a formalism for expressing security: a survey on theory and practice. ACM Comput. Surv. **55**(6), 1–36 (2023). https://doi.org/10.1145/3534967
3. Behrmann, G., David, A., Larsen, K.G.: A tutorial on UPPAAL. In: Bernardo, M., Corradini, F. (eds.) SFM-RT 2004. LNCS, vol. 3185, pp. 200–236. Springer, Heidelberg (2004). https://doi.org/10.1007/978-3-540-30080-9_7
4. Bodeveix, J.-P., Boudjadar, A., Filali, M.: An alternative definition for timed automata composition. In: Bultan, T., Hsiung, P.-A. (eds.) ATVA 2011. LNCS, vol. 6996, pp. 105–119. Springer, Heidelberg (2011). https://doi.org/10.1007/978-3-642-24372-1_9
5. Bunte, O., van Gool, L.C.M., Willemse, T.A.C.: On the preservation of properties when changing communication models. In: Gasieniec, L. (ed.) SOFSEM 2023. LNCS, vol. 13878, pp. 239–253. Springer, Heidelberg (2023). https://doi.org/10.1007/978-3-031-23101-8_16
6. Chevrou, F., Hurault, A., Quéinnec, P.: Automated verification of asynchronous communicating systems with TLA+. In: Electronic Communications of the EASST 72 (2015): Proceedings of the 15th International Workshop on Automated Verification of Critical Systems (AVoCS 2015), pp. 1–15. https://doi.org/10.14279/TUJ.ECEASST.72.1019
7. Chevrou, F., Hurault, A., Quéinnec, P.: On the diversity of asynchronous communication. In: Formal Aspects of Computing. Applicable Formal Methods 28.5 (Sept 2016): Papers on Formal Engineering Methods including Extended Versions of Papers Presented at ICFEM 2014. Part 2, pp. 847–879 https://doi.org/10.1007/s00165-016-0379-x. ISSN:0934-5043
8. Chevrou, F., et al.: A map of asynchronous communication models. In: Sekerinski, E., et al. (ed.) Formal Methods. Revised Selected Papers, Part II. FM 2019 International Workshops, Porto, 7–11 October 2019. LNCS, vol. 12233, pp. 307–322. Springer, Heidelberg (2019). https://doi.org/10.1007/978-3-030-54997-8_20
9. Clarkson, M.R., Schneider, F.B.: Hyperproperties. J. Comput. Secur. **18**(6) (2010). In: Sabelfeld, A. (ed.) 21st IEEE Computer Security Foundations Symposium (CSF 2008), pp. 1157–1210. https://doi.org/10.3233/JCS-2009-0393. ISSN:0926-227X

10. Gerking, C., et al.: Domain-specific model checking for cyber-physical systems. In: Famelis, M., et al. (eds.) Model-Driven Engineering, Verification and Validation. Proceedings of the 12th Workshop on Model-Driven Engineering, Verification and Validation. MoDeVVa 2015, Ottawa, 29 September 2015. CEUR Workshop Proceedings, vol. 1514, pp. 18–27 (2015). urn: urn:nbn:de:0074-1514-4
11. Gerlach, L.: Impact of asynchronous communication models on the composability of information flow security. Zenodo (2024). https://doi.org/10.5281/zenodo.11397611
12. Gerlach, L.: Untersuchung des Einflusses von Kommunikationsmodellen auf die Zusammensetzbarkeit von Informationsflusseigenschaften. German. BA thesis. Karlsruhe Institute of Technology (KIT) (2023)
13. Gerlach, L., Gerking, C.: Generic asynchronous communication modeling concepts. Zenodo (2024). https://doi.org/10.5281/zenodo.7967158
14. Di Giusto, C., et al.: A partial order view of message-passing communication models. In: Proceedings of the ACM on Programming Languages (POPL), vol. 7, no. 55, pp. 1601–1627 (2023). https://doi.org/10.1145/3571248
15. Goguen, J.A., Meseguer, J.: Security policies and security models. In: 1982 IEEE Symposium on Security and Privacy, Oakland, 26–28 April 1982, pp. 11–20. IEEE Computer Society (1982). https://doi.org/10.1109/SP.1982.10014. ISBN: 0-8186-0410-7
16. Lamport, L., Schneider, F.B.: Verifying hyperproperties with TLA. In: 34th IEEE Computer Security Foundations Symposium (CSF 2021), Dubrovnik, 21–25 June 2021, pp. 1–16. . IEEE (2021). https://doi.org/10.1109/CSF51468.2021.00012
17. Mantel, H.: Information flow and noninterference. In: van Tilborg, H.C.A.., Jajodia, S. (eds.) Encyclopedia of Cryptography and Security, pp. 605–607. Springer, Boston (2011). https://doi.org/10.1007/978-1-4419-5906-5_874
18. Mantel, H.: On the composition of secure systems. In: Proceedings of the 2002 IEEE Symposium on Security and Privacy, Berkeley, 12–15 May 2002, pp. 88–101. IEEE Computer Society (2002). https://doi.org/10.1109/SECPRI.2002.1004364. ISBN: 0-7695-1543-6
19. McCullough, D.: Noninterference and the composability of security properties. In: Proceedings of the 1988 IEEE Symposium on Security and Privacy, Oakland, 18–21 April 1988, pp. 177–186. IEEE Computer Society (1988). https://doi.org/10.1109/SECPRI.1988.8110. ISBN: 0-8186-0850-1
20. McCullough, D.: Specifications for multi-level security and a hook- up property. In: 1987 IEEE Symposium on Security and Privacy, Oakland, 27–29 April 1987, pp. 161–166. IEEE Computer Society (1987). https://doi.org/10.1109/SP.1987.10009. ISBN: 0-8186-0771-8
21. Parnas, D.L.: The secret history of information hiding. In: Broy, M., Denert, E. (eds.) Software Pioneers. Contributions to Software Engineering, pp. 398–409. Springer, Heidelberg (2002). https://doi.org/10.1007/978-3-642-59412-0_25
22. Sabelfeld, A.: The impact of synchronisation on secure information flow in concurrent programs. In: Bjørner, D., Broy, M., Zamulin, A.V. (eds.) Perspectives of System Informatics. Revised Papers. 4th International Andrei Ershov Memorial Conference. PSI 2001 (Akademgorodok, Novosibirsk, 2–6 July 2001. LNCS, vol. 2244, pp. 225–239. Springer, Heidelberg (2001). https://doi.org/10.1007/3-540-45575-2_22. ISBN: 3-540-43075-X
23. Szyperski, C.A.: Component software. Beyond object-oriented programming. In Collab. with Gruntz , D., Murer, S (eds.) 2nd edn. Addison-Wesley Component Software Series. Addison-Wesley, Longman (2002). http://www.worldcat.org/oclc/248041840. ISBN: 0-201-74572-0

24. Zakinthinos, A., Stewart Lee, E.: The composability of non-interference. J. Comput. Secur. **3**(4), 269–282 (1995). https://doi.org/10.3233/JCS-1994/1995-3404. ISSN:0926-227X

Author Index

D. Marmsoler and M. Sun (Eds.): FACS 2024, LNCS 15189, p. 147, 2024.
https://doi.org/10.1007/978-3-031-71261-6